A Double Image of the Double Helix

A Series of Books in Biology:
Cedric I. Davern, Editor

A Double Image of the Double Helix

The Recombinant-DNA Debate

Clifford Grobstein

University of California, San Diego

*The intricate hereditary crystal, turning in human hands,
can cast either a beneficent or a malevolent light.*

W. H. Freeman and Company

San Francisco

793764

Sponsoring Editor: Arthur C. Bartlett; *Project Editor:* Pearl C. Vapnek;
Manuscript Editor: Suzanne Lipsett; *Designer:* Marie Carluccio;
Production Coordinator: Chuck Pendergast; *Illustration Coordinator:* Batyah Janowski;
Artist: Evan L. Gillespie; *Compositor:* Graphic Typesetting Service;
Printer and Binder: The Maple-Vail Book Manufacturing Group.

Cover photograph: H. R. Wilson and M. H. F. Wilkins

Library of Congress Cataloging in Publication Data

Grobstein, Clifford, 1916–
    A double image of the double helix.

    (A Series of books in biology)
    Bibliography: p.
    Includex index.
    1.   Recombinant DNA — Social aspects.   I.   Title.
QH442.G77   574.8′732   78-26093
ISBN 0-7167-1056-0
ISBN 0-7167-1057-9 pbk.

Printed in the United States of America

9 8 7 6 5 4 3 2 1

Contents

CHAPTER SIX
The Issue of Research Regulation   83

EPILOGUE
Agenda for the Future   107

APPENDIXES

Preface

This book was written during much of the period it describes. Looking back over these several years, one can descry a kind of paroxysm of public concern over recombinant-DNA research, beginning in 1975, rising to a climax in early 1977, and subsiding in 1978. Now that the tumult and shouting have declined, one can take a calmer look at the behavior of the body politic in the face of a profoundly important issue that briefly but strongly agitated the public mind.

I am a developmental biologist, not a molecular geneticist. My active research career was in the field of embryonic development. Development and genetics are closely related but in the early part of this century the two diverged into separate disciplines. Progress in genetics moved more rapidly, as researchers pursued bold and astute assumptions by using the new tactics and techniques of molecular biology. As a result the DNA-RNA-protein trinity wrought by these molecular approaches has become the common base for greater understanding of both heredity and development.

I kept track of early molecular genetics because this neighboring research field clearly had implications for my own. Later, having turned to biomedical administration, I developed a great interest in the role of science in the genesis of public policy. In this context my attention refocused on the astonishing pace and profound potential of molecular genetics. Recombinant DNA, seen from this perspective, was not only a powerful biological tool but a precedent-making issue for effective public policy dealing with the generation of new knowledge. Two serious issues were raised: whether public health and welfare were threatened by recombinant-DNA research and, perhaps more basic, whether research itself needs greater public regulation.

This book, therefore, is an attempt to provide both scientific and public-policy perspectives on a major insight into basic life processes. Recombinant DNA and the thrust of molecular genetics, of which it is part, may fluctuate in public prominence but they will certainly continue to generate advances and opportunities — a certainty interpreted by some as challenging and by others as shattering. Tensions are inevitable when new knowledge creates dramatic and novel situations and potentials, and such tensions cannot be quickly or cleanly dispelled. We are reaching new levels of capability for biological intervention. Our own species, as well as those others allied with or inadvertently dependent upon us, cannot avoid being deeply affected. In this inevitability there are good grounds for both excitement and concern.

Passionate reactions, therefore, are understandable. The outcome of the controversy, however, must also involve a cooler, more rational appraisal. Both reason and passion appear in the record of the DNA issue, as they do in the history of all great events, for both factors mold decisions in human affairs, whether the decisions be made by individuals or societies.

I have, of course, my own passions and predilections, the products of a complex imprinting process that was itself shaped by my occupation as a scientist. These subjective factors are expressed in the treatment that follows. I have, however, tried to display other perspectives as well, along with the scientific facts that afford opportunities to achieve understanding. In controversial areas, I believe, those who have special factual knowledge should use it not as their own weapon but as a means to accommodate conflicting views. I confess to a preference for public policy that accommodates rather than eliminates conflicting views.

Such a public policy is not only a device to assure harmony. It is a process which necessarily acknowledges that the human course is in part an improvisation, something more than the playing out of a set program. Programs may be conceived in the minds of aggressive leaders, embodied in DNA, or cast in the rhetoric of divine revelation. Nonetheless, the course of human affairs is a complex resultant determined in no small part by the clash and accommodation of diverse predilections. From this process comes collective will, without which no public policy can be stable.

The recombinant-DNA issue has emphasized that each of us holds a different personal image of the future. Together time and the exercise of human will reduce the multiplicity of these images and lay down the actual track of history. Human will itself is a unique and increasingly powerful product and determinant of life's earthly progression. In an epochal transition human will now is extending beyond the earth. These great changes are not automatic; humans are making themselves to be less and less pawns and more and more masters of their fate. Therefore, we do well to ponder and even to agonize over each great new step. In this way we formulate our will, individually and collectively. Painful though the process may be, we can do no less. And someday, perhaps, through this process, we will learn how to do more.

August 1978   *Clifford Grobstein*

A Double Image of the Double Helix

CHAPTER ONE

The Path
to Asilomar:
The Long Prologue

Asilomar is a conference center close to the sea at the very tip of the Monterey Peninsula in central California. It is spectacularly beautiful. Nearby, the Pacific waves crash on fractured rocks and ebb and flow through crevasses and over clear tidepools. White sands pile up into dunes and Monterey cypress and pine are molded into strange shapes by the prevailing wind.

Little more than a stone's throw from the sea are the center's traditional longhouses and more modern and luxurious residential and conference buildings. These cluster under tall redwoods around a comfortable mess hall and an informal social center. The site was once a religious retreat; it is now operated by the State of California and made available to those who wish to take counsel together in the peaceful atmosphere, natural beauty, and relative solitude.

One building, close to the beach, stands somewhat alone. It is the former chapel. Though no longer ecclesiastical in function, the structure's architectural quality is still somehow spiritual, subdued and cloistered. Here, in February 1975, an international conclave of unusual character took place. One hundred and fifty scientists, of the variety known as molecular biologists, came together for three days to discuss not their science but the possible necessity for controlling it.

It was appropriate for them to gather together and fitting that they chose this place. The subject at hand was the chemistry of heredity, the material basis for the linkage of biological generations. These linkages are the substance not only of past evolution but of future biological destiny. The subject, the occasion, and the setting therefore were dramatic. The conferees were at a high pitch of creative excitement. They were also, however, apprehensive about possible dangers of their research and of public reactions to it. In particular they saw themselves as threatened by frustration. These scientists stood on the threshold of great new understanding, and yet they had been constrained to pause, at least temporarily. Each conferee had hopes and plans for rewarding new experiments. Yet, during the three cloistered days of the conference, the participants uneasily achieved a consensus that restraint and caution were necessary. Their consensus was unprecedented, and had implications they knew they could not fully forecast. Close to the unbounded and restless energy of the sea, in Asilomar's sanctuary-like setting, they agreed to recommend that their own freedom of action as researchers be limited, even though they had no certain proof that the need for limitation existed or that the consequences of it would be positive.

Why did the conferees come to this conclusion? To understand we must recall that scientists have in common a firm commitment to realism. Their primary objective is to comprehend and cope with a universe of external phenomena that they believe we all share. The personality and politics of scientists may vary, but realism is a common characteristic among them. Those who gathered at Asilomar knew that they and their colleagues in a dozen countries were breaking into a previously unexplored and fundamental aspect of real phenomena. They were certain that their success would have enormous potential for intellectual and social benefit. Yet some also foresaw possible danger. Among the conceivable dangers of research penetrating so deeply into heredity were adverse personal effects upon themselves, upon other people, and upon their own activities as scientists. It was a real problem in a real world and they had to confront it in some rational way.

Molecular biologists are scientists of a special kind, a breed that came into prominence only in the last quarter century. In less than three decades their fruitful efforts moved them into the forefront of biology, changed its orientation, and generated a "biological revolution." They are flushed with success, single-minded in the momentum of their continuing effort, and thoroughly occupied with the next steps to be taken. Yet a group of them was called to Asilomar to talk about what they could *not* do and who was going to tell them so. It was as though one of the long rollers of the neighboring ocean, rising to its crest and driven by inexorable inner energy, was suddenly told to cease moving and subside.

This book is concerned with analyzing the conference's reasons for being as well as the meaning and implications of its conclusions. Wrapped up in the story are profoundly important facts and issues. The subject of the Asilomar conference is a crucial example of new knowledge giving humans the power to intervene in and control the natural order. The power to intervene and control physical nature has raised many questions in the past that have never been fully answered. There is, however, a new dimension when intervention and control bear on life itself. By manipulating genetic substance could we inadvertently dislocate delicate balances and unleash catastrophe? What are our purposes, and are the risks appropriate to projected gains? Can we afford to forego a gain of knowledge that may prove to be another step in the long climb toward a still only dimly perceived human destiny?

Theme and Countertheme

The drama at Asilomar had a prologue at least a century long. The great theme of the prologue was the biological dictum that like begets like. Life is a complex, continuous fabric from its first firm establishment on earth some three billion or more years ago. Its continuity rests on the certainty that every living thing comes from a similar progenitor (see Figure 1-1). Bacteria give rise to bacteria, wheat to wheat, cats to cats, and dogs to dogs. No kind or species propagates except from predecessors of its own kind. This fundamental theme was definitively sounded by the experimental genius of Louis Pasteur. Pasteur saved the vintners of France by convincing them of the reality and consequence of microbial breeding in their fermentation vats. Simultaneously he laid low the notion of the spontaneous generation of life, the misconception that rodents or microbes could form *de novo* from miasmas, dung hills, or rotting meat.

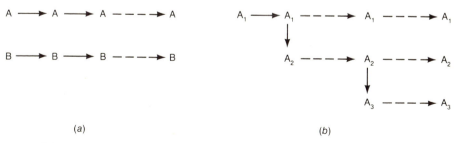

(a)   (b)

FIGURE 1-1

Hereditary Continuity and Variation in Generations of Organisms.
(a) Two organisms, A and B, continue to give rise to identical organisms indefinitely (dashed arrows).
(b) An organism, A, may propagate identical organisms indefinitely (top row) or vary to A_2 that propagates indefinitely (middle row) or again varies to A_3 (bottom row). Each *propagable* change is a hereditary variation.

The countertheme to the continuity principle that like begets like is the regular occurrence of exceptions — hereditary variation. Bacteria, wheat, cats, dogs, and all other kinds of living things vary in detail, and the properties of some of these variants are unlike those of any ancestor. These exceptions, once they arise, also propagate according to the theme of like begets like. The variants are not centaurs, Loch Ness monsters, or the charming creatures of Dr. Seuss. They are, for example, cats or dogs with different coat colors, of different size, or even with different patterns of behavior. With many complexities, the theme of like begets like and the countertheme of occasional hereditary variation are the central motifs of evolutionary biology, thrust into the consciousness of the late-nineteenth-century world by the English gentleman-naturalist Charles Darwin.

Chromosomes and the Basic Building Blocks

At the beginning of the present century it became clear that in higher organisms such as ourselves to a great degree heredity and variation are strongly linked to the behavior of the chromosomes that lie within the nucleus of cells. Gregor Mendel, the Austrian monk, had counted the hereditary variants among generations of garden peas and found their numbers to be transmitted in simple, intriguing ratios. These ratios, in turn, had their foundation in the regular dance of the chromosomes during cellular multiplication and particularly during the sexual formation of eggs and sperm to initiate a new generation. Every organism consists of one or more cells. In higher organisms all cells are descended by division from a first cell produced by the fusion of a sperm and egg derived from the preceding generation. The chromosomes of the fusion cells are derived from the chromosomes of the parents through sperm and egg. The chromosomes split in half time and again so that every descendant cell is provided with an identical set of chromosomes and essentially the same heredity (see Figure 1-2).

What in the chromosomes carries the heredity? Chromosomes are rich in both proteins and nucleic acids. These long threadlike molecules occur in combinations of enormous complexity that, a half-century ago, seemed to defy all understanding. However, by brilliant and patient splitting, picking, and sorting, in only two generations biologists have penetrated the complexity to discern an underlying principle of relative simplicity. Though there is still much to learn, the molecular basis for like begets like with occasional hereditary variation stands revealed.

The mystery of protein structure was solved first. It became clear that despite the enormous variety of proteins in existence, their properties stem from the sequence in each protein chain of only 20 different amino acid components (see Figure 1-3). Linked end to end by hundreds in specific sequences, the 20 kinds can provide endless diversity. For example, the first and second amino acids in a chain of 2 can each

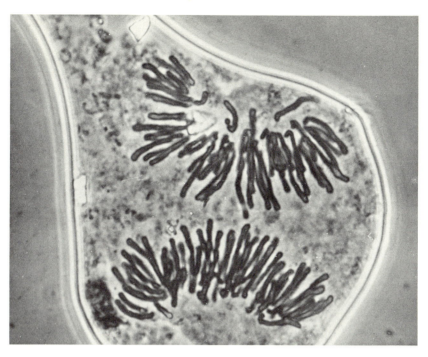

FIGURE **1-2**

Cell in Mitosis.
A root-tip cell of a hybrid lily (Black Beauty) undergoing replicative division. Identical chromosome sets have moved away from each other, so that each daughter cell will receive a complete set of hereditary instructions borne in the chromosomes. (Courtesy of Herbert Stern, University of California, San Diego.)

be any one of the 20. There are then 20 × 20 or 400 possibilities for a "chain" of two amino acids. Since each additional amino acid added to the chain multiplies the possibilities by 20, there are 8000 possibilities in three-amino-acid chains and 160,000 in four-amino-acid chains. Now consider that a chain of 100 amino acids is a *short* protein molecule. The enormously varied and complex proteins are based on a relatively simple principle of construction — 20 kinds of amino acids linked end-to-end in a fixed, specific sequence *peculiar to each protein.*

The variety and fixed sequence of proteins made them appear to early researchers to be good candidates as the carriers of heredity. However, though they proved to be importantly involved, proteins are not in fact the key. Carrying heredity from generation to generation is the role of the nucleic acids, and especially one of the two major kinds whose names are abbreviated as DNA and RNA. Deoxyribonucleic acid (DNA) has only four kinds of components, known as nucleotides (see Figure 1-4a). Like the amino acids in proteins, the nucleotides in nucleic acids are linked end to end and, also as in proteins, it is the sequence of components that is important. Even more crucial to the theme of like

5

Chemical "side-group"

Basic end $H_3N - C - C - OH$ Acid end
$\quad\quad\quad\quad | \quad \|$
$\quad\quad\quad\quad H \quad O$

(a) *A generalized amino acid*

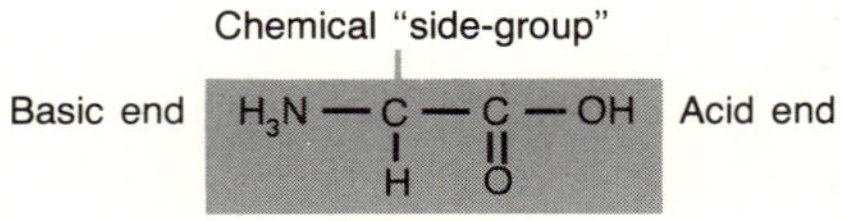

$CH_3 \quad CH_3$
$\quad\backslash\;/$
$\quad CH$
$\quad\quad |$
$H_3N - C - C - OH$
$\quad\quad | \quad \|$
$\quad\quad H \quad O$

Valine (val)

CH_3
$\;|$
S
$\;|$
CH_2
$\;|$
CH_2
$\;|$
$H_3N - C - C - OH$
$\quad\quad | \quad \|$
$\quad\quad H \quad O$

Methionine (met)

CH_3
$\;|$
CH_2
$\;|$
$H - C - CH_3$
$\quad\;|$
$H_3N - C - C - OH$
$\quad\quad | \quad \|$
$\quad\quad H \quad O$

Isoleucine (ile)

(b) *Three particular amino acids with different side-groups*

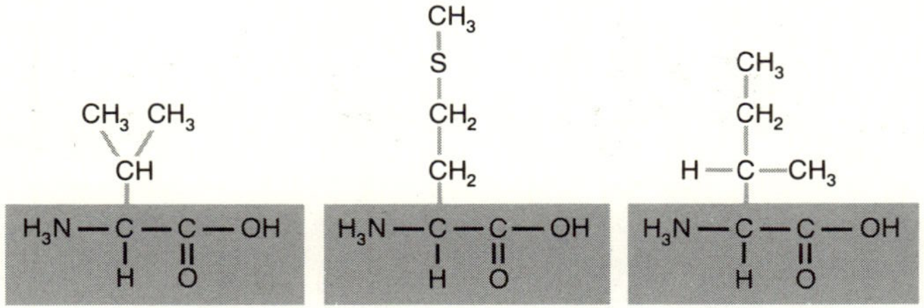

| val | met | ile |

(c) *A tripeptide*

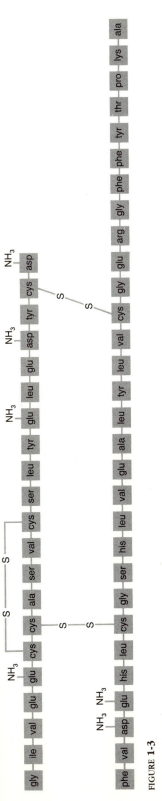

FIGURE 1-3

Amino Acids and Protein Structure.
Amino acids uniformly have basic and acid ends *(a)* that can link in peptide bonds to form chains *(c)*. *(b)* Each of 20 different amino acids has a different chemical "side-group." *(d)* Each protein has a specific and invariant amino acid sequence. The sequence and the side-groups determine how the chains couple and fold into three-dimensional configurations that underlie the actual properties of the proteins. Proteins serve essential functions, for example, as enzymes (catalysts), hormones, structural fibers, antibodies, and oxygen-carriers. Without every protein in its proper place and doing its proper job, the organism is handicapped or even inviable.

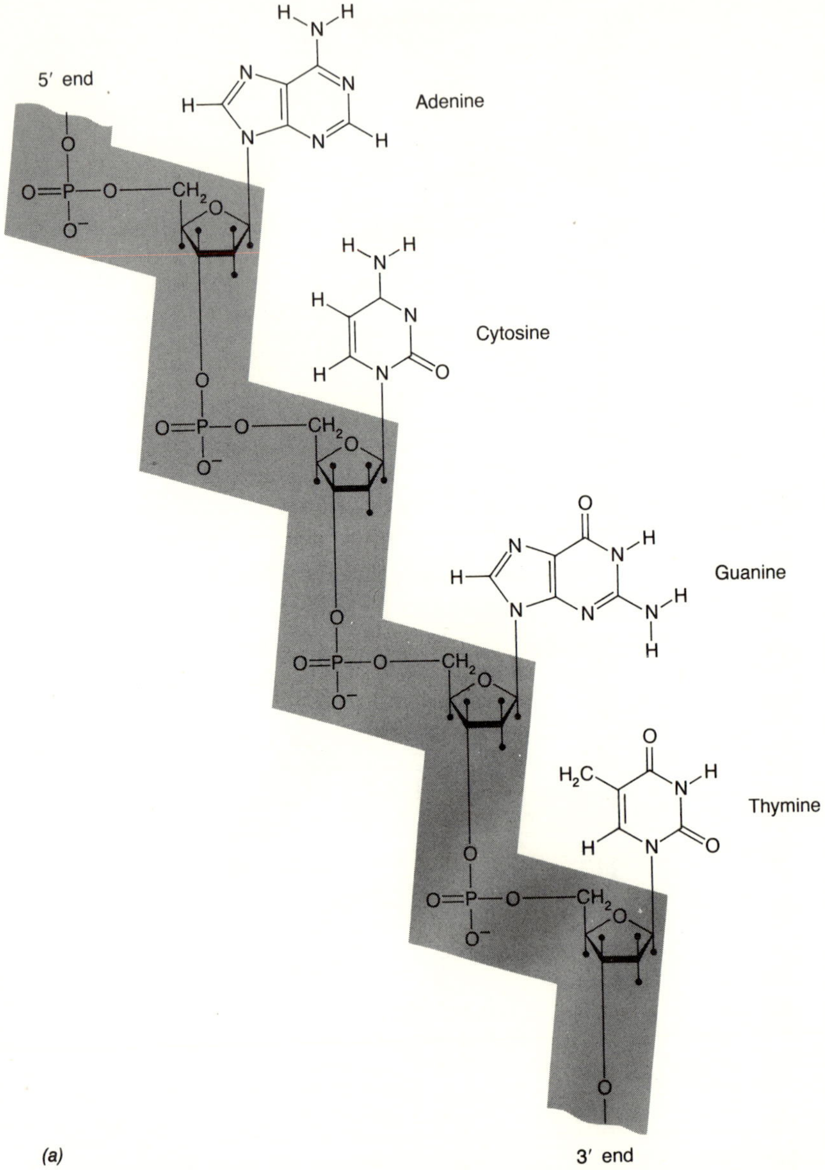

5′ end

Adenine

Cytosine

Guanine

Thymine

(a)

3′ end

FIGURE 1-4a

Nucleotides and the Double Helix of DNA.
The nucleotides of a single DNA strand are linked by ribose sugar to phosphate bonds
(shaded band). The four side-groups result in four kinds of nucleotides — adenine
(A), cytosine (C), guanine (G), and thymine (T).

begets like, however, is the unique ability of DNA, with the help of
proteins, to make exact copies of itself. This ability is part and parcel of
its remarkable but simple structure. When this structure was first proposed in 1953 by James Watson and Francis Crick almost every biologist

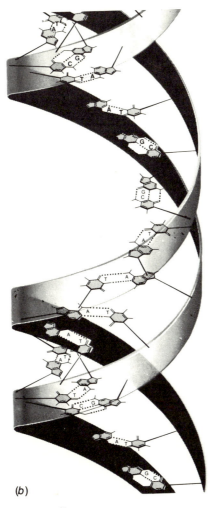

(b)

FIGURE **1-4b**

The double helix of DNA consists of two
coiled single strands (shaded bands)
held together by bonding between A
and T or between C and G. The two
strands thus are complementary and will
restore their pairing if separated.

was immediately struck by the beautifully simple way it explained much
of genetic regularity. Either the proposed structure of DNA had to be an
ingenious fantasy made up out of whole cloth to satisfy geneticists or
else the apparent enormous complications of heredity had, in one
stroke, been displaced by a profound and dramatic simplicity. The answer now is known. Watson and Crick were indeed ingenious, but the
double-helix structure of DNA that they proposed was no fantasy; it truly
exists as an "invention" of the real world of life phenomena.

The double helix consists of two interwound springlike chains of nucleotides that are joined end to end (see Figure 1-4b). The helical chains are interlinked by bridges (hydrogen bonds) between the component nucleotides. In each chain the nucleotides are in what appears to be random order. If we abbreviate the names of the nucleotides to A, C, G, and T, any one of these in the chain is followed by any other with about equal frequency, but in bridging between chains, A is always connected to T and G to C. The two interwound chains must therefore be complementary, a C in one always opposite to a G in the other, an A in one always opposite to a T in the other. Thus, a sequence -C-A-T- in one chain must be bridged to a sequence -G-T-A- in the other.

The complementary nature of the two strands has an important consequence (see Figure 1-5). The interwound and interlinked complementary strands form what is called double-stranded DNA. When they are unwound and unlinked, as happens both naturally and experimentally when certain changes occur in their chemical environment, they become two complementary single strands. In the presence of appropriate enzymes (protein catalysts) and abundant nucleotides of the four kinds, each single strand will form alongside itself a new bridged and complementary strand. Therefore, in effect, when double-stranded DNA is separated, each single strand reproduces or replicates the original double strand. Thus, the strand containing the sequence -C-A-T- assembles a complementary strand containing -G-T-A-. The result

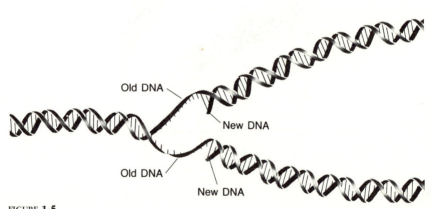

Old DNA

New DNA

Old DNA

New DNA

FIGURE 1-5

Replication of DNA.
Replication of DNA can be accomplished because of the complementary relation (A to T, C to G) between the nucleotide sequences on the two strands of the DNA molecule. Under appropriate chemical conditions, the bonds between the complementary bases are weakened and the two strands then can unwind and separate. In the presence of suitable enzymes and available nucleotides of the four kinds, a new chain will form by complementary bonding to the exposed nucleotides of each unpaired older chain. The new complementary sequence thus formed is then linked into a stable chain by an enzyme that promotes end-to-end nucleotide coupling. In this way two new helixes identical with the first are formed. (From "The Recombinant-DNA Debate" by Clifford Grobstein. Copyright © 1977 by Scientific American, Inc. All rights reserved.)

is two identical double strands where there was one before. Another
separation of strands doubles their number and results in four copies of
the original double strand. Each successive replicative cycle therefore
doubles the number of strands, and the process can continue indefinitely. Since the strands include genes, these hereditary units are also
duplicated in the process.

At the level of molecules, like begets like through DNA replication.
This phenomenon occurs in the reproduction of every organism on
earth and it has been happening, so far as we know, ever since the first
true life emerged eons ago. As a chemical process, the doubling of DNA
is undoubtedly the most prolific and portentous of nature's entire bag
of tricks.

Mutation and Its Replication

How does variation get introduced into this seemingly endless production of identical copies? As might be expected, through "error." Nothing
that happens so often can be totally free of mishap. Put another way,
nothing in nature occurs with 100 percent probability. In a certain
percentage of cases, mutational errors occur in the assembly of a new
strand (see Figure 1-6). With a frequency in the range of 1 in 100,000 to
10 million, the segment -C-A-T- may entirely fail to produce its com-

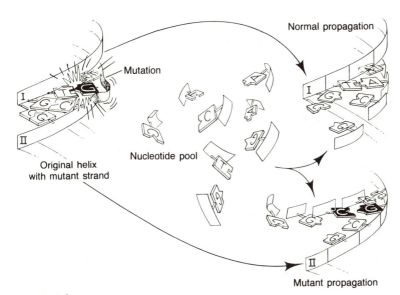

FIGURE **1-6**

Mutation Propagation.
Propagation of a mutation by "error" (substitution of G for A at left) in a
DNA strand (II). Once the error is made, replication may propagate it, as the
mistakenly placed G (identified as black) pairs with a C in the newly forming
strand. Thus the altered sequence may propagate as readily as the original one.

plementary -G-T-A-, or it may allow for example a C to get in where an A should be. In the first case the new strand will entirely lack the ability to produce -C-A-T- at the next separation. In the second case the new strand with -C-A-T- will make a -C-A-G- at the next separation. If we assume that the sequence -C-A-T- is necessary to make a normal cat, the first kind of error may mean that no cat can be produced. The new form would be "lethal" because it could produce no offspring. The second kind of error, however, may produce a new or mutant characteristic (-C-A-G-) of cats not seen in the parents but able to appear in subsequent offspring. If this variant is viable it may, either by nature or by fond cat-breeders, be incorporated into a new strain of cats. Such a change, of course, is very unlikely to produce a dog or any other completely un-catlike and yet viable creature. Many errors must accumulate in fortunate combinations to produce a species change.

What we have described is the basic molecular mechanism for heredity and variation. It rests entirely on the two intercoiled "springs" of DNA. They are the source of the remarkable reliability of reproductive replication with just the right dash of inevitable error. Between replication and mutation we get not only the spice of seemingly infinite human individuality but also the luxurious variety of living things that soften and embellish the face of the earth. Having seen the austere television images of the moon and Mars, we can appreciate the remarkable fecundity of the double springs of DNA out of which wells the rich variety of earthly life.

Molecular Genetics

The molecular concept of heredity emerged from the genetics of the fifties and was confirmed and elaborated in the sixties. The concept was achieved through a series of new techniques that combined the subtleties of controlled breeding and other genetic manipulation with powerful new tools provided by chemistry and physics. Most especially it depended upon a shift in the focus of investigation from higher organisms to bacteria and viruses. By this shift the "slow motion" of human or even fruitfly genetics was accelerated to a rate that better fits the time frame of experimental effort. Instead of human generations measured in years or fruitfly generations measured in weeks, generations in bacteria are measured in minutes and those in viruses in still shorter times. The short generation times make even rarely occurring genetic errors measurable and manipulable. Moreover, the small size and relative simplicity of bacteria and especially viruses allow direct chemical analysis of an abundance of identical organisms. What we know today of molecular genetics is mostly the product of studies of these simple living forms. We have just enough information about more complex organisms to be sure that they are not *entirely* different. They have not taken off on their own, but have built upon what the simple organisms "learned" during the perhaps 2 billion years they alone inhabited the earth.

The cells of simpler bacteria differ from those of higher organisms in not having a nucleus with a number of chromosomes. These cells of more primitive kind evolved before the advantages of a nucleus were "discovered." Such *procaryote* (prenuclear) cells have a single, very long, coiled but circular chromosome that is not separated from the rest of the cell by a membrane. They also frequently have smaller circular DNA molecules known as *plasmids*. The single procaryotic chromosome is much simpler than the multiple chromosomes of more complex organisms. The positions of various critical nucleotide sequences have been mapped on the bacterial chromosome and the relationships of many of these sequences to the activities and other characteristics of the bacterial cell have been worked out. We have grown increasingly certain that the sequence of the individual nucleotides in the chromosome is the basis for hereditary characteristics. Hence it has become more and more important to specify the sequences precisely and, if possible, to isolate, purify, and duplicate the key ones.

In the course of this research it was discovered that particular proteins are not only determined by nucleotide sequences in DNA but also are essential to DNA replication. Among other functions, proteins act as catalysts (enzymes) in linking nucleotides together in the construction of new DNA chains. Certain enzymes, called *nucleases,* also are active in splitting nucleotides apart. Some nucleases only break linkages between nucleotides that occur in particular sequences, for example -C-A-T-. A number of such nucleases, known as *restriction enzymes,* have been discovered, each targeted to break a different nucleotide sequence. These nucleases are used in the experimental fragmentation of long DNA chains into shorter segments whose ends are known because the breakage points can be controlled.

Restriction enzymes provided an added bonus to researchers. The enzyme-induced breakage of the strands does not always occur at exactly the same point on the two double strands (Figure 1-7a). Rather, the break on one strand may occur several nucleotides away from the break on the other, leaving a projecting unbridged end. Such ends are sticky toward sequences that are complementary (-C-A-T- will attach side to side with -G-T-A-). Thus, any two DNA fragments produced by a given restriction enzyme will rejoin because of complementary stickiness. An experimental technique is thereby provided for recombining DNA from different sources.

The recombination of DNA following breakage was by no means an unknown phenomenon. It had been detected decades ago in the breeding of complex organisms and it is a recognized mechanism of natural variation and evolution. It also was known to occur as an aspect of radiation damage and in various other genetic abnormalities. What was entirely novel was that when combined with a number of other experimental techniques this mechanism gave researchers the ability to pro-

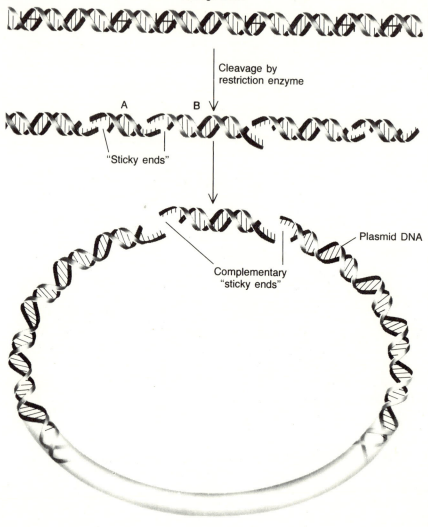

Foreign DNA

Cleavage by
restriction enzyme

A   B

"Sticky ends"

Complementary
"sticky ends"

Plasmid DNA

(a)

FIGURE 1-7a

Recombinant-DNA Technique.
DNA being cleaved and the fragments inserted into a plasmid. DNA (labeled
"foreign") is cleaved by a restriction enzyme to yield fragments A and B, each with
complementary "sticky ends." The same restriction enzyme is used to break open a
plasmid. Fragment B is shown about to be inserted to reclose the broken plasmid circle.
Fragment A might be inserted similarly to produce a second insert-bearing plasmid. (After
"The Recombinant-DNA Debate" by Clifford Grobstein. Copyright © 1977 by Scientific
American, Inc. All rights reserved.)

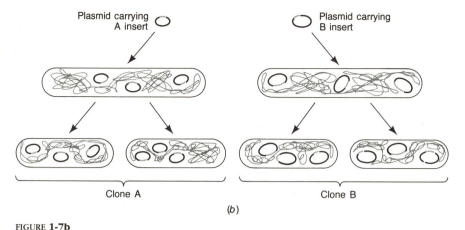

FIGURE **1-7b**

Two kinds of insert-bearing plasmids have been returned to *E. coli* cells, where they propagate along with the cells to produce two "clones."

duce controlled recombinations that would not normally be expected to occur in nature.

Moreover, these artificially combined (recombined) DNA molecules could be reinserted into bacterial hosts where they would replicate along with the host DNA (Figure 1-7b). This reinsertion could be done in two ways. One way took advantage of the fact that, as mentioned above, bacteria frequently have, in addition to their one large chromosome, a number of shorter, circular DNA molecules known as plasmids. Plasmids replicate independently of the chromosome, often confer resistance to antibiotics, and certain of them can be transferred from cell to cell—even of different bacteria—by a process known as *conjugation*. Also, they can be separated from the rest of the cell and from the larger chromosome. The isolated plasmids can then be treated with a suitable restriction enzyme to break them open and yield a linear DNA with sticky ends. When mixed with another DNA with complementary sticky ends, the plasmid DNA will combine with it. With further suitable manipulation a circular plasmid containing the foreign DNA is produced. After proper treatment, bacteria allow reentry of the recombinant plasmids and these can then reproduce indefinitely. Identical genetic replicas thus produced are known as *clones.*

The other method for reinsertion of recombinant DNA involves viruses known as bacteriophages. *Phages* propagate within bacterial cells and may eventually rupture them. On encountering new bacteria, the phage particles released by a ruptured cell can penetrate into them to start a new generation. Virus DNA can be isolated in the same way plasmid DNA can and similarly used for recombination with foreign DNA. The recombinants can then infect bacteria where the foreign DNA will propagate along with the viral and host DNA. In the language of molecular biology, the plasmid and viral carriers of foreign DNA are *vectors* and the organism that propagates the recombinant is a *host.* The host and its vectors, either viruses or plasmids, make up an experimental *system.*

It is the availability of such systems that has put the concept of genetic engineering — long recognized to be "down the road" — seemingly just over the next hill. The host for the system first discovered was the long-studied K – 12 strain of the bacterium *Escherichia coli* (*E. coli*). This common inhabitant of the lower intestine and rectum of humans and other organisms is by far the most common experimental subject of molecular genetics. It is the workhorse of virtually every molecular genetics laboratory, and its genetics is better known than that of any other species. It was particularly the ability to put specific DNA sequences from higher organisms into *E. coli* that excited investigators and produced grave apprehension in some. *E. coli* was the centerpiece of Asilomar and its concerns.

Beginnings of Concern

One of the early apprehensions that led toward the Asilomar conference is illustrated by the following story. The Cold Spring Harbor Laboratory has existed on Long Island in New York for half a century or more. It supports a relatively small but distinguished research program in genetics and is the site of an annual summer symposium that focuses on biological research at the forefront of the field. It also offers advanced courses each summer that bring together outstanding faculty and unusually promising students from institutions all over the country and abroad.

In the summer of 1971, Robert Pollack was an instructor in a course here in the techniques of mammalian cell culture, an area that underlies not only cell biology but studies of viruses, cancer, and development as well. In his last lecture of the course, Pollack turned from technique to broader matters of safety and ethics. He pointed to the need for precautions against immediate and known dangers and made reference to some of the less well-understood hazards of indiscriminately mixing cells of different species under artificial conditions. Some of his students were troubled by the warning. One pointed out that in Professor Paul Berg's laboratory at Stanford, from which she came, experiments were in progress to combine DNA from a virus and DNA from *E. coli* and then to insert the combination into the bacterium *E. coli* to propagate. The virus, SV40, was derived from mammals and could cause human cells in culture to take on a cancerlike appearance. The other *E. coli* DNA contained a biochemical "marker" so the combination could be followed when reintroduced into *E. coli*. The experiment suggested *E. coli* as a suitable host in which to study the action of genetic materials derived from other sources.

Pollack was astonished. The experiment described was indeed a

startling example of the concerns he was expressing. To put genetic material from an agent that at least mimicked certain properties of human cancer into a bacterium that normally inhabits the human gut seemed, to put it mildly, a trifle risky. Might the agent cause cancer if it accidentally got into the human gut? He telephoned Berg and explained his concerns. Berg listened patiently, then said that the experiment had not yet been done and that he would think about it.

Thus the concept of potential "biohazard" from recombinant DNA was born. Some six months later Berg called Pollack and asked him to join in organizing a conference on the safety and ethics of various lines of research then under way or planned. In the interval, Berg had wrestled with Pollack's questions, decided that his planned experiment should not be done, and reached the conclusion that new and serious issues of public health might be on the horizon. If so, these new issues would have to be more broadly addressed. A subsequent conference, in January 1973, led to a book, *Biohazards in Biological Research,** that elevated the issue in the consciousness of many biologists. Pollack had correctly sensed a problem, and Berg's amplification of Pollack's warning began the response that led to Asilomar. This kind of seemingly casual interaction and highly individualistic behavior is characteristic of the way the scientific community works. It illustrates the intense concentration of scientists on particular objectives, the wide net of their informal communication, and the effectiveness of what is called "peer review." The last is really collective judgment, and it is brought to bear both formally and informally. It underlies a good deal of the success of science.

Pollack was not the only person experiencing apprehension. Working at the Stanford Medical School in close vicinity to Berg was Stanley N. Cohen. Cohen's general objective was similar to Berg's but he and his collaborators were approaching the matter in another way. They took advantage of the existence in *E. coli* of plasmids. Cohen demonstrated in 1973 that the DNA of plasmids with different properties, suitably treated in test tubes, can be recombined and that the recombinants will display joint properties on return to the bacterial cell.

Within a matter of months, in 1973, the relatively simple procedure based upon plasmids was being used both in Cohen's and in a number of other laboratories to transfer into *E. coli* genetic materials from other microorganisms and from higher organisms. For example, Cohen and his colleagues found that bacterial-plasmid DNA will even recombine with DNA from toads, that it will carry the toad DNA back into the bacterial cell, and that the toad DNA will then replicate along with the plasmid. The technique spread quickly because Cohen made available to other investigators the plasmid he had found to be highly advantage-

*A. Hellman, M. N. Oxman, and R. Pollack, (eds.), *Biohazards in Biological Research* (Cold Spring Harbor, N.Y.: Cold Spring Harbor Laboratory, 1973).

ous for the process. However, he recognized the potential hazards of certain possible combinations and asked for assurance from his colleagues that no combinations would be made with tumor viruses or with any DNA that would make *E. coli* more resistant to antibiotics. He also asked his colleagues not to pass the plasmid on to others without informing him. This simple self-regulated control broke down when additional plasmids and mechanisms to accomplish DNA recombination were found in other laboratories. Cohen, like Berg, became convinced that a more general and effective mechanism to guard against possible risk was needed.

Spread of Concern

Meanwhile, the circle of concerned discussion had widened. Earlier, in June 1973, the annual Gordon Conference on Nucleic Acids had provided scientists with an opportunity to recognize the growing potential of the new technology. At the end of the meeting, during which Cohen described his early results with plasmids, a special session was called. The cochairperson of the Conference was Maxine Singer of the National Institutes of Health. She opened the special session with a sharply etched portrait of two emerging faces of DNA:*

We all share the excitement and enthusiasm of yesterday morning's speaker, who pointed out that the scientific developments reported then would permit interesting experiments involving the linking together of a variety of DNA molecules. The cause of the excitement is two-fold. First, there is our fascination with an evolving understanding of these amazing molecules and their biological action. Second, there is the idea that such manipulations may lead to useful tools for alleviation of human health problems. Nevertheless, we are all aware that such experiments raise moral and ethical issues because of the potential hazards such molecules may engender. . . . Because we are doing these experiments, and because we recognize the potential difficulties, we have a responsibility to concern ourselves with the safety of our coworkers and laboratory personnel, as well as with the safety of the public. We are asked this morning to consider this responsibility.

Following the morning discussion a letter was drafted to the presidents of the National Academy of Sciences and the Institute of Medicine asking that an *ad hoc* study group be constituted. The letter also was published in *Science,*† widely read in the scientific community. Its text follows:

We are writing to you, on behalf of a number of scientists, to communicate a matter of deep concern. Several of the scientific reports presented at this year's Gordon Research Conference on Nucleic Acids (June 11– 15, 1973, New Hampton, New Hampshire) indicated that we presently have the technical ability to join together, covalently, DNA molecules from diverse sources. Scientific

---

*See "Research with Recombinant DNA," *Academy Forum* (Washington, D.C.: National Academy of Sciences, 1977), p. 24.
†*Science* 181: 1114 *1973*.

developments over the past two years make it both reasonable and convenient

to generate overlapping sequence homologies ["sticky ends"] at the termini of different DNA molecules. The sequence homologies can then be used to combine the molecules by Watson-Crick hydrogen bonding [side to side]. Application of existing methods permits subsequent covalent linkage [end to end] of such molecules. This technique could be used, for example, to combine DNA from animal viruses with bacterial DNA, or DNA's of different viral origin might be so joined. In this way new kinds of hybrid plasmids or viruses, with biological activity of unpredictable nature, may eventually be created. These experiments offer exciting and interesting potential both for advancing knowledge of fundamental biological processes and for alleviation of human health problems.

Certain such hybrid molecules may prove hazardous to laboratory workers and to the public. Although no hazard has yet been established, prudence suggests that the potential hazard be seriously considered.

A majority of those attending the Conference voted to communicate their concern in this matter to you and to the President of the Institute of Medicine (to whom this letter is also being sent). The conferees suggested that the Academies establish a study committee to consider this problem and to recommend specific actions or guidelines, should that seem appropriate. Related problems such as the risks involved in current largescale preparation of animal viruses might also be considered.

Maxine Singer, National Institutes of Health

Dieter Soll, Department of Molecular Biophysics and Biochemistry, Yale University

Paul Berg of Stanford University, a member of the National Academy, was asked by the Academy President in February 1974 to chair the suggested study committee. Thus began consideration of appropriate public policy to deal with recombinant DNA research and uses.

Formal Deliberations

When the Berg committee convened in April 1974, the group agreed immediately to recommend that an international conference be called early the following year. The committee was aware of the successful but still unpublished transfer of toad genes into *E. coli*. They foresaw that research pressure was building up so rapidly, moreover, that it would be necessary to take some action pending the conference. Accordingly, the committee drafted a second letter to be published in *Science*. The letter briefly outlined the committee's concerns and asked that "scientists throughout the world join with the members of this committee in voluntarily deferring" certain types of experiments and carefully weighing others. The committee also requested that the Director of the National Institutes of Health establish an advisory committee with specific charges to deal with the rapidly developing situation. This letter is so important in making clear the thinking of involved scientists at the time that it is reproduced here in full:*

Science 185: 303 *1974.*

Recent advances in techniques for the isolation and rejoining of segments of DNA now permit construction of biologically active recombinant DNA molecules in vitro. For example, DNA restriction endonucleases, which generate DNA fragments containing cohesive ends especially suitable for rejoining, have been used to create new types of biologically functional bacterial plasmids carrying antibiotic resistance markers and to link *Xenopus laevis* ribosomal DNA to DNA from a bacterial plasmid. This latter recombinant plasmid has been shown to replicate stably in *Escherichia coli* where it synthesizes RNA that is complementary to *X. laevis* ribosomal DNA. Similarly, segments of *Drosophila* chromosomal DNA have been incorporated into both plasmid and bacteriophage DNA's to yield hybrid molecules that can infect and replicate in *E. coli*.

Several groups of scientists are now planning to use this technology to create recombinant DNA's from a variety of other viral, animal, and bacterial sources. Although such experiments are likely to facilitate the solution of important theoretical and practical biological problems, they would also result in the creation of novel types of infectious DNA elements whose biological properties cannot be completely predicted in advance.

There is serious concern that some of these artificial recombinant DNA molecules could prove biologically hazardous. One potential hazard in current experiments derives from the need to use a bacterium like *E. coli* to clone the recombinant DNA molecules and to amplify their number. Strains of *E. coli* commonly reside in the human intestinal tract, and they are capable of exchanging genetic information with other types of bacteria, some of which are pathogenic to man. Thus, new DNA elements introduced into *E. coli* might possibly become widely disseminated among human, bacterial, plant, or animal populations with unpredictable effects.

Concern for these emerging capabilities was raised by scientists attending the 1973 Gordon Research Conference on Nucleic Acids, who requested that the National Academy of Sciences give consideration to these matters. The undersigned members of a committee, acting on behalf of and with the endorsement of the Assembly of Life Sciences of the National Research Council on this matter, propose the following recommendations.

First, and most important, that until the potential hazards of such recombinant DNA molecules have been better evaluated or until adequate methods are developed for preventing their spread, scientists throughout the world join with the members of this committee in voluntarily deferring the following types of experiments:

Type 1: Construction of new, autonomously replicating bacterial plasmids that might result in the introduction of genetic determinants for antibiotic resistance or bacterial toxin formation into bacterial strains that do not at present carry such determinants; or construction of new bacterial plasmids containing combinations of resistance to clinically useful antibiotics unless plasmids containing such combinations of antibiotic resistance determinants already exist in nature.

Type 2: Linkage of all or segments of the DNA's from oncogenic [tumor-causing] or other animal viruses to autonomously replicating DNA elements such as bacterial plasmids or other viral DNA's. Such recombinant DNA molecules might be more easily disseminated to bacterial populations in humans and other species, and thus possibly increase the incidence of cancer or other diseases.

Second, plans to link fragments of animal DNA's to bacterial plasmid DNA or bacteriophage DNA should be carefully weighed in light of the fact that many types of animal cell DNA's contain sequences common to RNA tumor viruses. Since joining of any foreign DNA to a DNA replication system creates new recombinant DNA molecules whose biological properties cannot be predicted with certainty, such experiments should not be undertaken lightly.

Third, the director of the National Institutes of Health is requested to give immediate consideration to establishing an advisory committee charged with (i) overseeing an experimental program to evaluate the potential biological and ecological hazards of the above types of recombinant DNA molecules; (ii) developing procedures which will minimize the spread of such molecules within human and other populations; and (iii) devising guidelines to be followed by investigators working with potentially hazardous recombinant DNA molecules.

Fourth, an international meeting of involved scientists from all over the world should be convened early in the coming year to review scientific progress in this area and to further discuss appropriate ways to deal with the potential biohazards of recombinant DNA molecules.

The above recommendations are made with the realization (i) that our concern is based on judgments of potential rather than demonstrated risk since there are few available experimental data on the hazards of such DNA molecules and (ii) that adherence to our major recommendations will entail postponement or possibly abandonment of certain types of scientifically worthwhile experiments. Moreover, we are aware of many theoretical and practical difficulties involved in evaluating the human hazards of such recombinant DNA molecules. Nonetheless, our concern for the possible unfortunate consequences of indiscriminate application of these techniques motivates us to urge all scientists working in this area to join us in agreeing not to initiate experiments of types 1 and 2 above until attempts have been made to evaluate the hazards and some resolution of the outstanding questions has been achieved.

Paul Berg, *Chairman;* David Baltimore; Herbert W. Boyer; Stanley N. Cohen; Ronald W. Davis; David S. Hogness; Daniel Nathans; Richard Roblin; James D. Watson; Sherman Weissman; Norton D. Zinder.

Committee on Recombinant DNA Molecules Assembly of Life Sciences; National Research Council, National Academy of Sciences, Washington, D.C. 20418.

Toward Asilomar

The stage was now set for the Asilomar Conference. The immediate excitant for the conference could be called the dawning of a new perspective. Something like this had happened some forty years earlier when nuclear physicists of the thirties were tracking the nature of the inner binding forces of the atom. Suddenly some saw the potential for an immensely expanded scale of energy release, conceivably useful in achieving a new level of military destructiveness—the atom bomb. Similarly, molecular geneticists of the early seventies were tracking the inner intricacies of hereditary transmission. Suddenly some saw the possibility of inadvertent or deliberate production of new lethal disease. The two experiences make clear that knowledge is as close to intervention as imagination is to myth—whether for good or evil. The seeker of "truth" singlemindedly concentrates on expanding knowledge. He or she usually neither sees, speaks, nor plots good or evil. But new objective truth, when refracted in a multifaceted social structure, can emerge with conflicting shapes and implications.

J. R. Oppenheimer was the scientist who organized the social translation of fission physics into the atom bomb. Oppenheimer was also a

social philosopher. He observed poignantly after the first atomic explo-
sion that "physicists have known sin; and this is a knowledge which they
cannot lose." President Harry Truman said of the same event, "I casually
mentioned to Stalin that we had a new weapon of unusual destructive
force. The Russian Premier showed no special interest. All he said was
that he was glad to hear it and hoped we would make good use of it
against the Japanese."

Oppenheimer's "sin" seemingly was a mere implement in the
complex calculus of international politics. At Asilomar, molecular
geneticists first confronted a new double image of fundamental knowledge, in this instance the double image of the double DNA helix. The
marvellous crystal of hereditary information, turning in human hands,
clearly could cast a malevolent as well as a beneficent light.

CHAPTER TWO

First Approximation: Self-Regulation

How does one begin a conference on a subject that the conferees are not sure they even want to discuss? How does one plan an agenda to accomplish a purpose still to be defined? Can 150 people of strong mind arrive at a consensus when they almost certainly have very different views and are not sure they share objectives? These were the questions faced by Paul Berg, Chairman of the Asilomar conference. In reply to a question of procedure, David Baltimore, who made the opening statement of the conference, answered, "The procedures by which the consensus will be determined will be largely determined by the extent of the consensus." In short, the only way to proceed was to wade in and see what would happen.

The conferees had come together from all over the world. That they had come at all, and almost everyone invited did, meant that they could not ignore the issues raised. Many, however, had come because they were afraid to allow decisions to be made in their absence. Their own work was a piece of the problem to be discussed and could be inhibited or even terminated by whatever transpired. Not many of them believed that the experiments they themselves planned to do were dangerous. They knew, however, that the agencies who supported them and the politicians who appropriated funds for the supporting agencies might see danger and intervene in their work, perhaps delaying or otherwise inhibiting their efforts. Therefore, they came, not to bury Caesar but to defend him.

The Minimalists and the Moderates

Many of the participants were "minimalists." They hoped for as little restriction of research as possible — by anybody. The minimalists were amenable to collective action, but preferred that such action should bolster and even certify continuance of research, especially their own. Most of the minimalists would have liked to avoid the matter altogether but feared that if they and their conferees didn't do something, someone else might — to their disadvantage. After all, the National Academy of Science and the National Institutes of Health had already been asked to address the matter. Above all the minimalists feared legislative regulation in language that would ensnare them in endless lengths of red tape. They didn't, however, entirely trust their colleagues as regulators either. Among the minimalists were youngsters not yet trained to any bridle. They had allies among the elders, many of whom had no taste for bridles because they had lived their professional lives in open pastures where the direction taken by research was absolutely unrestricted.

However, there was also a significant group of moderates. These were people who had already wrestled with their doubts and concerns and had come to accept the necessity, however regrettable, for some kind of agreed-upon controls. They hoped, however, to meet the need through self-regulation within their own community. They knew that external regulation was a real possibility, and they sought to minimize it by demonstrating that involved investigators could themselves take responsible collective action. They disagreed as to the extent of the risk and also because they had different ideas about what might satisfy outside concerns. The organizers of the conference were themselves in this moderate group. Even they, however, were not of one mind as to the stringency of regulation that their colleagues would find necessary or acceptable. Their lack of consensus over details and their uncertainty as to how their colleagues would react accounted for the absence of a preformulated agenda of specific activities designed to meet established objectives.

There were no maximalists at the conference. No one came to say *mea culpa*, I have sinned, judge and punish me. No one wanted to argue that recombinant-DNA research should be permanently banned and all already published papers burned. No one wanted to establish a high tribunal to decide what experiments could be done, by whom, and when. Such ideas are held by very few scientists. Any scientist who might have held such views would have sought a more receptive audience than the one at Asilomar. No one came to the conference to speak

for more regulation or inhibition of research than was clearly necessary to control manifest health hazards or to mitigate external control. By agreement, no one raised social or ethical questions. It was generally felt that such issues were outside the business of this assemblage and that dispute over them would only be futile and waste time.

The main conference debate centered on possible variants of the moderate position. The minimalists challenged the moderates repeatedly in the open and even more frequently in *sotto voce* asides to their neighbors. However, the inability of the minimalists to disclaim any possibility of risk allowed the conference to shift steadily toward the moderate position, just as preconference developments had moved the organizers to call the conference itself. The conference got down to specifics when the group dealing with problems posed by plasmids presented a classification of risk at 6 levels and suggested a graded series of procedures to contain them. From that point on, though much else was said and debated, the significant work of the conference dealt with defining risk and providing balancing containment.

The issues thereby became technical, familiar, and subject to rational debate and negotiation. The minimalists exerted themselves to keep the risk categories as general as possible, to avoid establishing rigid procedures for circumstances that still were not clearly defined. The moderates argued that sufficient detail had to be provided to assure the outside world (everyone not at the conference, but particularly legislators and officials who would exert control) that the product of the conference would not be a self-serving facade behind which responsible and irresponsible researchers alike could do as they wished.

The Concept of Containment

This interplay led to a reduction of the categories of risk from six to four. It also amplified the concept of containment in ways that made consensus easier. A key disagreement centered on the wisdom of the continued use of the $K-12$ strain of *E. coli*. Its enormous advantage as the best-known subject for genetic study had to be balanced against several other properties. This strain of *E. coli* was derived from progenitors with the ability to infect humans. Its ecological relationships were relatively unknown. And doubts existed as to the degree to which any disease-causing recombinant that might be produced from it could be physically contained.

Introduction of the concept of *biological containment* — as distinct from physical containment — gave greater maneuverability and eased the tensions over containment considerably. The concept involves the use in recombinant-DNA experiments of organisms so modified genetically that they are unable to survive and spread under natural conditions. These organisms might be strains of *E. coli* specifically modified to require conditions for survival not found in nature, or they might be organisms other than *E. coli* that are limited to very special and confined natural ecological niches. Coupling biological containment with physical containment would improve the estimated effectiveness of total containment. It would also make possible the practical postponement of certain types of experiments without prohibiting them, since development of the necessary techniques for biological containment still lay in the future. The time necessary for constructing facilities to provide effective physical containment, as well as that needed to develop organisms that would assure biological containment delayed the time when conceivably dangerous experiments could otherwise be done.

Self-Regulation Versus External Regulation

In the background to the formal discussion the moderates were exerting continuous pressure. At meals, between sessions, and late at night they argued, explained, and cajoled, making the case that some form of regulation was inevitable and that self-regulation was preferable to external regulation. The case was fortified by a session with four invited lawyers. The effect of this meeting was comparable to that of reality knocking on the door after the murder scene in Macbeth. The lawyers gave a background briefing on such aspects of the real world as the legal responsibility of supervisors for employee injury, the theory of legal responsibility, torts, the right of legislatures to be wrong in their interpretation of the public welfare, the ABCs of the requirements for occupational health and safety, the financial liability of institutions, and the wide dissemination and high costs of liability insurance. This briefing took place the night before the final session. The substance of the discussion was not the usual fare for molecular biologists. It shocked the conferees considerably and made them more willing to accept even inconvenient self-regulation rather than risk the rough waters of legislative and judicial restraint.

The conference organizers had spent much of the night before the final session distilling and reworking drafts provided by panels and the earlier plenary discussions. They had produced a five-page preliminary statement incorporating the concept of balanced risk and containment. They had fleshed out the statement with additional ideas, some frankly their own and some based on their interpretation of the consensus. It was a compromise but it was a whole plan, bringing order to the diffuse and somewhat chaotic discussion. The statement surprised some conferees by its definitiveness and frustrated others with its vagueness.

The organizers' paper was a calculated effort to place the bridle between the conference's jaws without causing it to rear. There was no doubt that it had regulatory content; many conferees could estimate the delays it would impose on their work and the expenditures of research funds for installing the safety equipment it would require for the experiments they had in mind. On the other hand, the statement prohibited almost nothing that was immediately in the offing, given that investigators were able and willing to make the necessary preparations. Doing so would take time and trouble, but there were no insurmountable obstacles. With respect to implementation the paper was mercifully vague. It referred to the fact that "national bodies" were even then formulating "codes of practice for the conduct of experiments with known or potential biohazard." Pending these codes it provided a "guide" and placed responsibility solely in the hands of individual investigators.

Still there was some flinching and protesting. One or another sentence or paragraph bothered one or another conferee. How were such matters to be resolved? The final chips were now going down; how would a consensus be judged? At this point the conferees wanted a vote. They overruled the organizing committee's strategy of a judged consensus with the committee as judge. The document was taken up section by section and on each section there was a show of hands. To the surprise of many, on no section were there more than a half-dozen dissenters. The document was approved overwhelmingly as a whole. Despite many individual disagreements and much confusion and debate a consensus had been achieved. The active and conflicting debaters had been a small percentage of the conferees. The "silent majority," as one of the organizers noted, listened and then voted solidly moderate. The organizing committee had read the situation correctly and done its job well.

The Asilomar Consensus

To recapitulate, the involved investigators who were called together at Asilomar accepted the fact that there was significant but not yet definable and measurable potential risk in certain types of recombinant-DNA experiments. They therefore recognized a requirement for regulation to minimize identifiable risk as much as possible. They strongly supported meeting this requirement through collective self-regulation and they sought to make a responsible start in that direction. They recognized the possibility of external regulation and sought to ameliorate it. They did not, however, seriously or specifically consider either the justification for or the objections to external regulation, and gave little thought to the mechanisms that might be involved.

The summary statement of the Asilomar Conference is a historical document and is reprinted in this book as Appendix I. Its first section contains a statement that was never seriously debated: "that most of the work on construction of recombinant DNA molecules should proceed. . . ." There follows the proviso that *was* extensively debated, not in principle but as to detail: "that appropriate safeguards, principally biological and physical barriers adequate to contain the newly created organisms, [should be] employed." Moreover, "certain experiments . . . ought not to be done with presently available containment facilities."

In the section on principles the basic strategy is enunciated: "the effectiveness of the containment should match, as closely as possible, the estimated risk." The hope is expressed that "within and between the nations of the world, the way in which potential biohazards and levels of containment are matched would be consistent." Reliance for containment was placed primarily on "biological barriers" involving, as one means, organisms unable to survive and propagate except in the laboratory. "Physical containment . . . provides an additional factor of safety." Different means of containment "will complement one another" so that improvements in biological containment can "permit modifications of the complementary physical containment requirements."

There follows a section containing recommendations for matching types of containment with types of experiments. These are "broadly conceived and meant to provide provisional guidelines for investigators and agencies concerned with research on recombinant DNA's." The types of containment are classified as to estimated risk: minimal, low, moderate, and high. The nature of the risk and the kind of physical containment are defined in general terms. A broad characterization of types of experiment is provided along with a statement of the physical and biological containment requirements for each.

The fourth section is on implementation. It emphasizes again the importance attached to biological containment and to research on new procedures of this kind. It also emphasizes skillful and effective laboratory technique and the need for educational programs for laboratory personnel. Continual assessment of potential hazard as knowledge and experience grow is viewed as essential. The final section stresses the gaps in existing knowledge for evaluating various hazards and urges the planning and carrying out of research in several directions to fill these gaps. It is especially urged that the gaps be filled "before large scale applications of the use of recombinant DNA molecules is attempted."

Translating the Consensus

The conferees carried the summary statement from Asilomar back to their countries of origin. In each country the subsequent events were different but in virtually all countries represented at the conference steps were initiated toward safety regulation of the general kind agreed to at Asilomar.* In the United States the summary statement went im-

---

*See "Report of the Federal Interagency Committee on Recombinant DNA Research: International Activities," November 1977. Available from the Associate Director for Program Planning and Evaluation, National Institutes of Health, Bethesda, MD, 20014. The summary and Part A of this report are reprinted as Appendix VI in this volume.

mediately to the Program Advisory Committee on Recombinant DNA Molecules (heretofore referred to as the Recombinant Advisory Committee), earlier appointed by NIH Director Donald S. Frederickson and chaired by NIH Associate Director DeWitt Stetten. This committee promptly ran into trouble in its attempt to translate the Asilomar generalizations into concrete terms. A first draft by a subcommittee was regarded by the whole committee as too stringent, but a second draft was felt to be too lenient by Paul Berg and some fifty petitioning biologists, most of whom had not been at Asilomar. In response, Chairman Stetten appointed a new subcommittee to produce a third draft. The committee itself met to consider all three drafts and hammered out a consensus in a two–day meeting at La Jolla, California early in December 1975. This meeting again dealt largely with how to classify various kinds of experiments with respect to required containment. By now it was clear that these specifics of containment were really the focus of contention among the molecular geneticists. Many of them were committed to particular lines of investigation which, if placed one notch higher in containment requirements, could be put out of reach in dollars (to modify facilities) or time (to develop new strains of organism or to shift to safer organisms). The argumentation was tough, passionate, and not entirely free of personal interest.

The final committee version was judged to be at least as stringent as the Asilomar consensus and probably more so. It went to the NIH Director Donald Frederickson, who had now been made aware that not only was there significant contention within the molecular biology community but also critics were appearing who challenged the whole Asilomar approach. The director was under two other kinds of pressure as well. Molecular biologists all over the world, still respecting the self-imposed moratorium, were by now growing restive at the further delay, and the possibility existed that unlimited experimentation might resume even without guidelines. Second, there were signals that molecular biologists abroad might oppose the adoption in their own countries of regulations more stringent than those drafted at La Jolla. International standardization might then be impossible.

To afford opportunity for testimony to those not yet officially on record, the director set up a special hearing before his own general advisory committee. Included at the hearing were David L. Bazelon, chief judge of the District of Columbia Court of Appeals; Peter B. Hutt, former general counsel of the Food and Drug Administration; and Philip Handler, President of the National Academy of Sciences. Debate centered largely on the balance of risk and containment in the guidelines, with critics asserting that the guidelines were inadequate to contain the risk, and defenders insisting that the requirements already imposed in the guidelines would delay research and would be found, with more knowledge, to be unnecessary.

Robert Sinsheimer, Chairman of the Biology Department at the California Institute of Technology, was among the earliest and severest critics of the emerging guidelines.* He argued that the very act of recombining DNA from bacteria and higher organisms risks unknown dangers. Since bacteria and higher organisms diverged in evolution a very long time ago, they clearly have evolved very different mechanisms of genetic expression. To now combine these mechanisms through experimental intervention could pose unknown but conceivably incalculable hazards to the whole world ecosystem. Moreover, Sinsheimer argued, because of the replicative power of new recombinants, releasing them into the environment would be essentially irreversible. The consequences, therefore, might be unprecedented not only in magnitude but in kind. To avoid these enormous dangers, he urged, all further recombinant-DNA research should be confined to a single, maximum-containment facility, at least until more information becomes available. Sinsheimer further questioned the right and wisdom of any human intervention in evolutionary processes, asking whether the human species is ready to make such momentous decisions. Others joined in voicing his partly technical and partly ethical and social concerns, and the latter became a significant but minor theme in sessions otherwise largely devoted to technical issues of biological and health hazard.

The cogency of the critics' arguments led Judge Bazelon to recommend to the NIH director that in the interest of public information he carefully disclose his reasoning on all issues that he might later resolve. Director Frederickson did just that in the guidelines he issued on June 23, 1976. He supplied a special "decision document" (Appendix II) preceding the actual guidelines (Appendix III). Here he made clear that in considering the comments derived from the meeting of his own advisory committee he referred some to the Recombinant Advisory Committee and received their reaction and advice after their meeting of April 1976. On the basis of advice and comments from these and other sources, he reached decisions on the various issues presented. The resulting guidelines are thus those formulated by the Recombinant Advisory Committee to carry out the Asilomar consensus, modified by the director only as to issues raised by critics, primarily at or subsequent to the hearing conducted before the Director's Advisory Committee.

The NIH Guidelines

The guidelines themselves provide a regulatory framework for recombinant-DNA research supported by the NIH. The introduction to the guidelines states the problem: "to construct guidelines that allow the promise of the methodology to be realized while advocating the

---

* R. L. Sinsheimer, "Recombinant DNA—on our own," *Bioscience,* 1 26 October *1976:*599; "Troubled dawn for genetic engineering," *New Scientist* October 16, *1975:*148–51; "Potential risks," *Academy Forum* (Washington, D.C.: National Academy of Sciences, 1977), pp. 78–80.

considerable caution that is demanded by what we and others view as potential hazards." It also repeats and somewhat amplifies the basic strategy of risk versus containment worked out at Asilomar. It then discusses containment, pointing out, as was done in the Asilomar statement, that standard microbiological practices, special physical containment equipment, and special biological procedures either existing or to be developed make up the conceivable containment resources. Each of these is described in detail (see Table 2-1). Physical containment is defined at four levels. The minimal level (P– 1) is that of standard microbiological practices, the low level (P– 2) provides a minimal degree of separation from surrounding facilities and somewhat more stringent laboratory procedures, the moderate level (P– 3) requires special engineering and safety equipment to reduce exchange of air, waste products, and personnel with surrounding areas, and the high level (P– 4) is a specially constructed facility in which everything possible is done to avoid inadvertent transfer of hazardous materials or organisms to the outside.

Biological containment is discussed in relation to possible kinds of experiments done. First, a set of experiments is designated as potentially too hazardous to be performed at the present time, given the existing state of knowledge. Permissible experiments are then outlined under three headings of increasing containment. The first involves minimal biological containment, and includes "most of the currently available systems" for recombinant-DNA experimentation. For example, it was judged not certain but likely that the K– 12 strain of E. coli provides some biological containment because it has been relatively enfeebled for survival in natural habitats, including the intestine and rectum of man, by long laboratory manipulation. K– 12 and its associated vectors therefore are designated as having EK– 1* biological containment value. The second level of containment (EK– 2) is provided by specially constructed or selected strains that are tested by laboratory experiment and are judged to provide effective containment. High containment is defined as an escape and survival frequency no greater than one in a hundred million. The EK– 2 classification can only be made by the Recombinant Advisory Committee from actual test data submitted to it. The third level of containment (EK– 3) is similar to the second but requires test of survival rates "in animals, including humans or primates, and in other relevant environments." This classification also must be certified by the Recombinant Advisory Committee.

Specific classes of experiments are then discussed as to both physical and biological containment requirements. These include (1) experiments using the E. coli K– 12 systems currently most in use; (2) new bacterial systems expected to be developed and to have containment advantages over E. coli K– 12; and (3) systems involving cells of higher organisms, both plant and animal. The treatment attempts to cover all

*E. coli, K– 12.

TABLE 2-1

Containment Requirements of the National Institutes of Health Guidelines for Recombinant-DNA Research (From "The Recombinant-DNA Debate" by Clifford Grobstein. Copyright © 1977 by Scientific American, Inc. All rights reserved.)

<table>
<tr><td rowspan="2" colspan="2"></td><td colspan="3" align="center">BIOLOGICAL CONTAINMENT (FOR E. COLI HOST SYSTEMS ONLY)</td></tr>
<tr><td align="center">EK1</td><td align="center">EK2</td><td align="center">EK3</td></tr>
<tr>
<td rowspan="2">CONTAINMENT</td>
<td>P1</td>
<td>DNA from nonpathogenic prokaryotes that naturally exchange genes with E. coli

Plasmid or bacteriophage DNA from host cells that naturally exchange genes with E. coli. (If plasmid or bacteriophage genome contains harmful genes or if DNA segment is less than 99 percent pure and characterized higher levels of containment are required.)</td>
<td></td>
<td></td>
</tr>
<tr>
<td>P2</td>
<td>DNA from embryonic or germ-line cells of cold-blooded vertebrates

DNA from other cold-blooded animals and lower eukaryotes (except insects maintained in the laboratory for fewer than 10 generations)

DNA from plants (except plants containing known pathogens or producing known toxins)

DNA from low-risk pathogenic prokaryotes that naturally exchange genes with E. coli

Organelle DNA from nonprimate eukaryotes. (For organelle DNA that is less than 99 percent pure higher levels of containment are required.)</td>
<td>DNA from nonembryonic cold-blooded vertebrates

DNA from moderate-risk pathogenic prokaryotes that naturally exchange genes with E. coli

DNA from nonpathogenic prokaryotes that do not naturally exchange genes with E. coli

DNA from plant viruses

Organelle DNA from primates. (For organelle DNA that is less than 99 percent pure higher levels of containment are required.)

Plasmid or bacteriophage DNA from host cells that do not naturally exchange genes with E. coli. (If there is a risk that recombinant will increase pathogenicity or ecological potential of host, higher levels of containment are required.)</td>
<td></td>
</tr>
</table>

Physical containment	"SHOTGUN" EXPERIMENTS USING *E. COLI* K–12 OR ITS DERIVATIVES AS THE HOST CELL AND PLASMIDS, BACTERIOPHAGES OR OTHER VIRUSES AS THE CLONING VECTORS (INDICATED IN ROMAN TYPE).	*EXPERIMENTS IN WHICH PURE, CHARACTERIZED "FOREIGN" GENES CARRIED BY PLASMIDS, BACTERIOPHAGES OR OTHER VIRUSES ARE CLONED IN E. COLI K–12 OR ITS DERIVATIVES (INDICATED IN ITALIC TYPE).*
	DNA from embryonic primate-tissue or germ-line cells	DNA from nonembryonic primate tissue
	DNA from other mammalian cells	*DNA from animal viruses (if cloned DNA contains harmful genes)*
	DNA from birds	
	DNA from embryonic, nonembryonic or germ-line vertebrate cells (if vertebrate produces a toxin)	
	DNA from moderate-risk pathogenic prokaryotes that do not naturally exchange genes with *E. coli*	
	DNA from animal viruses (if cloned DNA does not contain harmful genes)	
P3	DNA from nonpathogenic prokaryotes that do not naturally exchange genes with *E. coli*	DNA from nonembryonic primate tissue
	DNA from plant viruses	*DNA from animal viruses (if cloned DNA contains harmful genes)*
	Plasmid or bacteriophage DNA from host cells that do not naturally exchange genes with E. coli. (If there is a risk that recombinant will increase pathogenicity or ecological potential of host, higher levels of containment are required).	

Some examples of the physical and biological containment requirements set forth in the NIH guidelines for research involving recombinant-DNA molecules, issued in June, 1976, are given in this table. The guidelines, which replaced the partial moratorium that limited such research for the preceding two years, are based on "worst case" estimates of the potential risks associated with various classes of recombinant-DNA experiments. Certain experiments are banned, such as those involving DNA from known high-risk pathogens; other experiments, such as those involving DNA from organisms that are known to exchange genes with *E. coli* in nature, require only the safeguards of good laboratory practice (physical-containment level *P1*) and the use of the standard *K–12* laboratory strain of *E. coli* (biological-containment level *EK1*). Between these extremes the NIH guidelines prescribe appropriate combinations of increasing physical and biological containment for increasing levels of estimated risk. (In this table containment increases from upper left to lower right.) Thus physical-containment levels *P2*, *P3*, and *P4* correspond respectively to minimum isolation, moderate isolation, and maximum isolation. Biological-containment level *EK2* refers to the use of new "crippled" strains of *K–12* incorporating various genetic defects designed to make the cells' survival outside of laboratory conditions essentially impossible. Level *EK3* is reserved for an *EK2*-level host-vector system that has successfully passed additional field-testing. Experiments with animal-virus host systems (currently only the polyoma and SV40 viruses) require either the *P3* or the *P4* level of physical containment. Experiments with plant-virus host systems have special physical-containment requirements that are analogous to the *P1*-to-*P4* system.

experiments now envisioned, but, as noted in the introduction to the guidelines, "cannot hope to anticipate all ... lines of imaginative research that are possible with this powerful new methodology." When one considers the vast array of kinds of organisms on the face of the earth, the number of possible recombinations, in practical terms, is infinite. Given the power of the methodology plus the variety of possibilities, the guidelines for experimentation are necessarily directed at only a small sample of the experiments that will be conceived as knowledge grows.

The guidelines assign responsibility for continuing surveillance and decision to three sites: the individual investigator, the sponsoring institution, and the NIH. The investigator is assigned "primary responsibilities" and these are spelled out in nineteen categories. The institution is assigned the same responsibilities as the investigator, but these are to be "fulfilled on its behalf by the principal investigator." This stipulation takes into account the formal situation that the NIH makes grants to the institution as an administrative unit but on behalf of the investigator as the scientific decision maker. In addition, however, the institution is to establish a biohazards committee of "experience and expertise" to advise on policies and review and approve to the NIH the "facilities, procedures, and practices, and the training and expertise of the personnel involved" in applications to the NIH for recombinant-DNA research support. This committee is assigned two other responsibilities: to make public the membership and minutes of the committee and to ensure "acceptability of its findings in terms of applicable laws, regulations, standards of practices, community attitudes, and health and environmental considerations."

The NIH has a tripartite role in overseeing the relevant research. Its initial review groups, known as study-sections, are to review applications involving recombinant DNA research not only for scientific merit but also for biohazard, containment precautions, and new problems that the study-section itself cannot resolve. The latter are to be referred to the Recombinant Advisory Committee, established to advise the HEW Secretary, Assistant HEW Secretary, and NIH Director on broad technical problems connected with the program. The NIH staff is assigned five specific administrative responsibilities to assure smooth operation of the guidelines. The NIH as a whole has eleven specifically described responsibilities, including the five assigned to staff.

Under this division of responsibilities, individual investigators monitored by their peers are relied on heavily for the safe conduct of research, both by the sponsoring institution and the NIH. Scientific policy-making is made a peer function in the Recombinant Advisory Committee acting through the NIH and HEW. Other aspects of policy fall within the purview of the institutional biohazards committee and the NIH Director. As a federal official, the latter is responsible for seeing that all research complies with the regulations of the Environmental Protection Agency, the Occupational Health and Safety Administration, and all other relevant federal regulatory agencies. Ultimately the whole process

depends on the approval and support of the President and the Congress. The effort of the Asilomar conference to achieve primarily self-regulation is preserved in the guidelines but the involvement of NIH places self-regulation within a framework into which grades of external regulation can be readily introduced.

The guidelines conclude with footnotes and a set of appendixes providing technical data on alternatives to *E. coli* and its vectors as a test system, and on physical containment facilities, equipment, and procedures. These details express the purpose of the guidelines: to translate the Asilomar consensus into practical terms. The controversies within the Recombinant Advisory Committee and most of what surfaced at the director's special hearing in February 1976, had to do with how to spell out further the generalizations made at Asilomar. The guidelines are the means to achieve the objectives agreed upon there; they do not violate what is acceptable to the molecular biology community as a whole. Stirrings of concern beyond that community remained unaddressed, although they were becoming apparent and were taken note of in the director's decision document. By the time the guidelines were issued, new public theatres of discussion and decision were beginning to be established. A major question raised in the wider public discussion was: Given a capability for self-regulation by the research community through the NIH, how much external discussion and regulation is appropriate and required? We turn to this broader question in a later chapter.

CHAPTER THREE

Exploration and Risk

Why use recombinant-DNA techniques if they may involve conceivable risk? Why particularly if the risk can be conceived to be very high? One possible answer is that the conceivable benefit is very large and the risk-benefit ratio favorable. Risk-benefit analysis of this kind is very difficult to apply to recombinant-DNA research because both the risk and benefit currently are uncertain quantities.

There is another kind of answer, unquantifiable but stubbornly defended by explorers, whether they climb, navigate, or lead intellectual adventures. It is the answer that motivated some of our most inspiring human achievements—for example, the exploration of the oceans, the continents, the moon, and now, as we enlarge our scope technologically, our neighboring planets. There are always unknowns in such explorations. Risk is rarely absent but it is never considered a decisive obstacle. Explorers proceed *despite* risk; they are even invigorated by it. They do not ignore risk and danger, but instead are challenged by the necessity to plan carefully to minimize it. In these terms, recombinant-DNA research is a new and powerful way to explore and the uncharted territory to be explored is the operation of hereditary messages in such complex organisms as humans.

It is important to emphasize that those who developed recombinant-DNA techniques were on this path when the policy debate over their research erupted. They were actually a small party in a large expedition. The main body of molecular geneticists had successfully explored some of the deepest questions confronted by twentieth-century biology. In a quarter-century they had discovered the profoundly meaningful structure of DNA and verified that it was the central thread of hereditary continuity. They had demonstrated that duplications of DNA occurs regularly as a direct consequence of its structure, both within dividing cells and in cell-free chemical reactions in test tubes. They had shown that single strands of DNA, when placed in a reaction vessel containing nucleotides as building blocks and appropriate enzyme proteins as catalysts, instruct the formation of conplementary DNA copies. And they had found that DNA, in addition to serving as a program or template for *replication* of DNA copies, also provides a program for what is called *transcription* of a second kind of nucleic acid, RNA (ribonucleic acid).

Such RNA copies (see Figure 3-1) also are complementary to their corresponding DNA, but in a slightly different nucleotide "language." In turn, this *messenger RNA* can serve as the program for *"translation"* of its sequence of nucleotides into a corresponding sequence of amino acids in the synthesis of proteins. The translation involves a fixed *genetic code* in which a sequence of three nucleotides (*codon*) is "read" by a complex mechanism that selects and positions a particular amino acid. The translation process requires a number of accessory proteins as enzymes, but its consequence is a specific sequence of amino acids that, through two steps (DNA to RNA, RNA to protein) is always determined by the original DNA nucleotide sequence. This is crucial because, in its turn, the amino acid sequence establishes the character, shape, and other functionally important properties of the protein. Thus DNA, RNA, and protein are tightly linked as a triad that passes replicated hereditary DNA sequence to protein sequence and thus to the characteristics of each succeeding generation.

To convey the vast biological significance of this molecular genetic concept is not easy. Comparisons can be misleading, but one is tempted to say that the hereditary protein-nucleic acid triad is comparable in understanding of biological processes to such physical concepts as gravitation. Consider how it has clarified the older hereditary notions of genotype and phenotype. This hard-won distinction had been achieved at the turn of the century through ingenious breeding experiments with organisms that had been carefully selected for their special experimental advantages. These experiments revealed that the actual results of sexual reproduction could not always be predicted from the visible characteristics of the parents. Genotype therefore was conceived as a kind of hereditary abstraction. It might or might not be expressed in parents, but nonetheless was passed to offspring as a collection of dis-

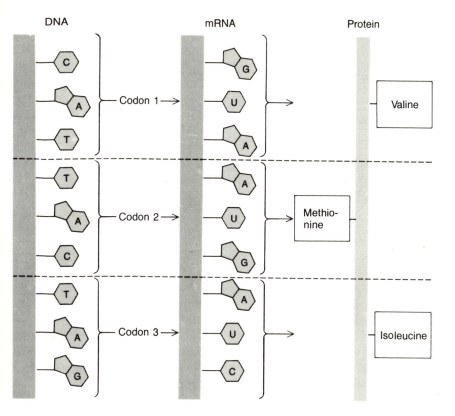

FIGURE 3-1

Sequence Relations Between DNA, RNA, and Protein.
The role of DNA in protein synthesis is shown in this highly schematic diagram. The nucleotide sequence CATTACTAG contains triplet "codons" CAT, TAC, and TAG. Each codon determines a complementary sequence in messenger ribonucleic acid (mRNA). In turn the mRNA codon positions a specific amino acid in a protein molecule. Therefore, the sequence of nucleotides in a given DNA molecule specifies a corresponding sequence of amino acids in a particular protein. The sequence of amino acids in turn establishes both the structure and the function of the protein. The letter "U" stands for the pyrimidine uracil, a constituent of RNA corresponding to thymidine in DNA. (From "The Recombinant-DNA Debate" by Clifford Grobstein. Copyright © 1977 by Scientific American, Inc. All rights reserved.)

crete and sometimes silent "factors" that could be displayed as characteristics in the offspring or in their subsequent breeding behavior. Genotype was established as residing in chromosomes, but its exact material nature and its manner of expression or nonexpression were unknown. Descriptively, for example, two normal parents that produced an albino offspring (without pigmentary coloring) were said to carry a genotypic factor for albinism. The factor was not expressed in either parent but was expressed when their genotypes combined in the offspring. The parental *pheno*type therefore was not albino even though the parents carried a covert *geno*type factor. Phenotype was the actual property (albino *vs.* nonalbino); genotype was the hereditarily transmitted potential, sometimes expressed and sometimes not.

If the foregoing seems obscure, note how the distinction became crystal clear when the new molecular genetic concept revealed genotype to be tangible, isolable, and manipulable DNA. The resulting concept asserted that phenotype, of whatever final complexity, begins with the formation of DNA-instructed proteins and that these closely controlled translations of genes further construct and control the course of development. Genes — specific and concrete elements of the genotype — thereby became hereditarily transmitted nucleotide sequences that were not abstractions but chemical information to regulate the operations of cells.

The New Vista

Once this molecular genetic vantage point was reached a whole new vista became apparent. One possible subject for exploration was the mechanism controlling the replication of DNA. Another was the mechanisms of its transcription to RNA. A third was to understand the manner in which translation of RNA to proteins occurs. Clearly, answers to some very old and important questions were now ripe for attack. For example, how does transmitted hereditary constitution (replicated DNA sequence) get transformed in detail into the delicately balanced complexities of structure and function that we see in adult organisms? How does the seeming simplicity of a fertilized egg, containing combined parental genotypes, become transmuted into a chicken, a mouse, or a human?

Much had been and would continue to be learned with respect to these questions through study of the genetics of *E. coli*. But the dominant problem was the nature of the genetic system of complex organisms and especially of humans. Human structure and function clearly involve complexity far beyond anything seen in *E. coli*. Human genetic organization must almost certainly have corresponding complexity as well.

For example, *E. coli* has a single circular chromosome, but humans have 46 chromosomes in 23 pairs. Each pair is unique, and each chromosome is composed not only of nucleic acids but of many kinds of proteins. Moreover, the bulk of the chromosomal DNA of complex organisms, unlike *E. coli*, consists of many identical copies of particular sequences. Only a very small fraction of the DNA consists of "unique" sequences occurring only once, as is the general rule in *E. coli* DNA. These special features of complex genetic systems could never be understood by studying simple ones alone. However, the combination of genetic and chemical techniques that proved so successful for *E. coli* cannot easily be applied to higher organisms. In these organisms direct

genetic analysis is necessarily slow and difficult because their generation time is long. It was exactly this limitation that researchers had
overcome earlier by turning to bacteria and viruses as experimental
objects.

For these reasons investigators plotting an experimental assault on
the genetic structure (*genome*) of such complex organisms as
humans — dreamed of riding at least part of the way on the reliable and
convenient *E. coli*. They knew that any alternative procedure meant
learning entirely new approaches and at least long delays. The strategy
of using *E. coli* became especially promising when it was found that
carefully fragmented DNA from complex organisms was not fundamentally different from *E. coli* DNA fragments. *E. coli* fragments could be
recombined with plasmid or virus DNA and reinserted into *E. coli* to
yield insight into how the *E. coli* genome is put together and how it
functions. Why should the same procedure not be applied to more
complex genomes?

Techniques for purifying DNA from complex organisms of other
chromosomal materials had been perfected earlier. As already noted,
such DNA differed from that of *E. coli* in that a large part of it consisted
of many copies of the same sequences. The multiple-copy sequences
could be separated from the smaller fraction that consisted of single or
unique sequences. The unique-copy portion appeared to correspond
with the *E. coli* DNA, which is all single-copy. Why should this portion
not be inserted into *E. coli* to provide a first step toward understanding
how DNA of complex organisms is organized and controlled?

As an example, genes of *E. coli* that contribute to several steps in a
specific biochemical process are frequently adjacent in the chromosome. Such a group of genes is called an *operon* (see Figure 3-2). There

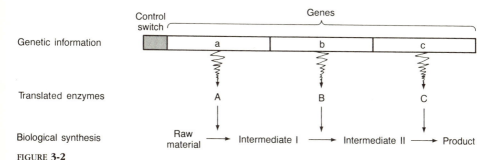

FIGURE **3-2**

An "Operon" — Multiple Genes Producing Multiple Enzymes Necessary for Making Biological Products.
In the biological synthesis shown (bottom row), a raw material (nutrient) is converted to a
necessary product through two intermediate stages. Three enzymes (A, B, C) are required to catalyze
(accelerate) the three steps. Each enzyme is determined in its highly specific properties by a gene
(a, b, c). The three genes are adjacent in the chromosome and have a single "control switch"
that turns on the genes to make enzymes when the appropriate raw material is present. When the
product is made in excess, it may act on the control switch to turn off the genes. The set of genes
and the single switch that controls them make up an operon — well known in bacteria but
hypothetical in higher organisms.

is, for example, an operon for utilizing the sugar, lactose, as a source of energy. When lactose is available as a nutrient source the lactose operon provides the required information for making several proteins necessary to decompose lactose and release its energy. However, if no lactose is available, it is inefficient. Accordingly, the lactose operon is "turned off" until lactose becomes available again. Having the genes involved in lactose breakdown arranged in a close cluster makes it easier for one "switch" to turn them on and off together. To accomplish this, blocks of contiguous *E. coli* genes have, at the end of the nucleotide sequences that code for the necessary proteins, "control" sequences that start and stop the genic expression at the appropriate signal.

This knowledge of *E. coli* is derived from the application of various recombination techniques. Suppose, now, that functional pieces of DNA from a complex organism could be individually inserted into *E. coli* cells as recombinants. Suppose that each such piece propagated along with the *E. coli* DNA to make many copies, thus resulting in a "library" of functional sections of the mammalian genome, each in a propagating *E. coli* strain. *E. coli*-type genetic techniques and chemical procedures could then be used to study the behavior of the inserted DNA segments. By putting together various pieces of higher-organism DNA in the rapidly reproducing *E. coli*, investigators might detect interactions that would give clues to the normal organization of complex hereditary systems.

Exploring with a Shotgun

This is bold speculation and experimentation stemming from the normal exercise of scientific ingenuity. This kind of ingenuity has been spectacularly successful in penetrating the complexities of heredity and many other natural "mysteries." It represents the way scientific investigation picks at and gradually dispels the unknown. Problems arise, however, in this specific instance. The tactic of fragmenting a complex and poorly understood genome and inserting its sections randomly into *E. coli* has been dubbed a "shotgun experiment" (see Figure 3-3). Briefly, a shotgun experiment breaks an unknown DNA into segments and indiscriminately introduces them into *E. coli* hosts. The various hosts, each containing a piece of the genome under study, is then dispersed on an appropriate nutrient base. Each individual recombinant cell divides and grows to form a separate colony. As mentioned earlier, such hereditarily uniform colonies descended from a single progenitor are called clones. Each colony, containing a particular recombinant DNA, provides an abundant cloned and therefore pure source of a particular DNA segment to be studied. The DNA segments now can be identified both chemically and genetically. Theoretically, with all of the segments eventually identified, their essential interactions as a total genome could be reconstructed.

The practical problems that arise in a shotgun experiment stem from the efficiency of the method for obtaining and amplifying particular DNA sequences. The approach is sophisticated and powerful, but its very power raises possible risk. Power and risk are often two faces of the same coin. Suppose that each cloned DNA segment were an actual gene of the complex donor organism and expressed itself in its *E. coli* host by forming its coded protein. This is not certain to happen and now seems unlikely. Yet the possibility had initially to be assessed, since it is one kind of objective for doing such an experiment in the first place. Suppose, further that the activity of one of these cloned genes normally

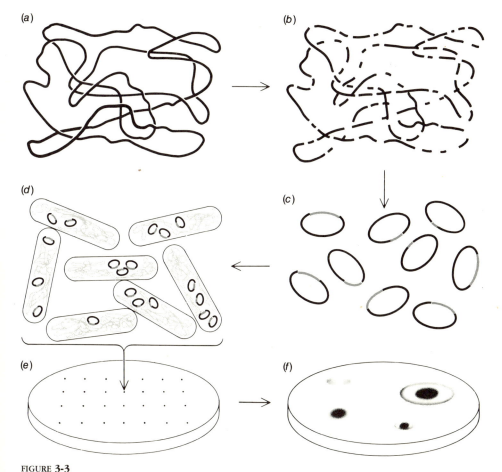

FIGURE **3-3**

The Stages in a Shotgun Experiment.
In a "shotgun" experiment, the total DNA of an organism *(a)* is exposed to restriction enzymes in order to yield many fragments *(b)*, which are then recombined with the DNA from a suitable vector *(c)* and randomly reinserted with the vector into the host cells *(d)*. The *E. coli* hosts are next spread on a nutrient substrate *(e)*, so that each recipient cell, containing a particular inserted foreign nucleotide sequence, can grow into a colony *(f)*. (From "The Recombinant-DNA Debate" by Clifford Grobstein. Copyright © 1977 by Scientific American, Inc. All rights reserved.)

is inhibited by the activity of another in the intact genome. When the controlled and controlling genes are separated the inhibition might be removed and large amounts of the coded protein might be produced. Suppose also that the produced protein is toxic or cancer-producing in the original species and that recombinant bacteria carrying the uncontrolled gene are accidentally spilled in the laboratory and infect the experimenter. Could the recombinant strain then act as a new pathogen created in the laboratory experiment?

To evaluate this possibility without more information is not easy. The less precedent we have the harder predicting the outcome of the exploration will be. In the untrammeled world of pure science unprecedented experiments are especially interesting for just this reason. But the more potent the possible *practical* consequences the greater the anxiety aroused by uncertainty. Since no one can say without some experience what an untested recombinant DNA from a complex organism will do, the possible existence of a deleterious effect clearly cannot be ignored. Moreover, it seems likely that the possibility will vary considerably depending on the particular recombinant DNA segment involved. Over the whole range of a given genetic constitution certain sections are probably more likely and others extremely unlikely to yield dangerous effects.

In a shotgun experiment many different DNA segments of unknown function are involved. Therefore, making a general estimate of the possible danger is impossible. The dilemma illustrates how deeply molecular genetics has now penetrated into fundamental aspects of all life. Further study, even in pursuit of seemingly purely theoretical objectives, *cannot avoid raising questions of practical consequence.* This is the central issue raised by recombinant-DNA research. It presents vast opportunity for enhanced understanding and control of heredity in complex organisms, but simultaneously and unavoidably it presents a set of practical risks of uncertain magnitude. Are these uncertain risks to be treated as surmountable obstacles to the course of exploration? Or are they so unique in magnitude and character that they define a *terra incognita,* an area too dangerous to be safely explored by seekers after knowledge?

To confront this fundamental issue — risk in relation to exploration — we need to restate it more precisely. What specifically is it that poses risk? How great is the risk in each particular instance? What steps can be taken to minimize the risk? In short, we need to make as complete an assessment as possible of the nature of the risk-obstacle before us and the possible means available to challenge it. Conceivably, the obstacle we face may be too difficult to overcome at the present time. It is equally conceivable that the obstacle may look menacing only because its true nature is obscured and distorted by our ignorance. Therefore, calm and careful consideration is needed. In particular, we must separate irrational dread or superstitious concerns about what we may find from the real world risks involved.

Shotgun experiments do maximize uncertainty. Suppose we choose to explore the genetic constitution of mice by the shotgun technique. No one knows for sure how many genes there are in a mouse, but 100,000 is within the range of estimates. Suppose 1 in 1000 of these genes has a significant probability of producing a pathogenic or otherwise deleterious strain of *E. coli*. On testing the mouse genome 1 gene at a time, the chances of producing a harmful strain in the first 10 recombinations would be only 1 percent. One might regard that risk as reasonable, assuming that all precautions are taken until greater knowledge is gained. However, if the shotgun approach is used, 100 harmful strains will be produced immediately, assuming that all 100,000 mouse DNA sequences are successfully inserted. We will not know in advance which strains are among the 100. Moreover, if we add the fact that the DNA obtained from the mice may contain DNA of parasitic viruses, bacteria, or other species, the number of harmful strains may be even higher. Viral DNA sequences in particular might be suspect because harmful strains are known to exist in "silent" state in chromosomes of healthy mice but to be activated on release.

Given these uncertainties, why carry out a shotgun experiment? The answer is, because without this technique isolating and testing individual mouse genes would require extremely laborious procedures that only work for a few special genes. We face the situation of the chicken and the egg: We cannot isolate more genes until we better understand the genome, and we will only better understand the genome by isolating more genes. The rationale of the shotgun experiment is to break the circle by wholesale cloning of unknown genes for later identification. To gain this advantage, however, some risk of dealing with the unknown must be taken.

How great actually is the risk? There can be no precise answer without exploratory experiments. Even the earlier assumption we made that 1 in 1,000 recombinants might be deleterious is entirely arbitrary. We might, with equal justification, have assumed 0, 1 in 100, or 1 in 10,000. Indeed, it has even been seriously suggested that *every* such recombinant could be dangerous. We will return to the rationale for this suggestion in later discussion.

In using shotgun procedures to continue the exploration, therefore, we must accept the fact that although warnings of extreme danger admittedly are speculative they can neither be disregarded nor effectively evaluated without more information. Can we sufficiently minimize risk while at the same time assessing it? The alleged danger is the production of an organism *through whose propagation and spread unfortunate consequences may occur.* Unlike the production of a chemical or nuclear explosive, therefore, the threat is not immediate and local but requires time and the transmission of an agent over some distance. Therefore, effective containment *within the experimental environment* could control the risk.

The need for containment is *not* an unprecedented circumstance or requirement. It is, in fact, the situation that gave rise to our highly successful technology for the control of infectious disease. Containment begins with technical proficiency achieved through careful training and quality control achieved through enforcement of professional standards and discipline. Further containment can be provided either physically or biologically. As noted earlier, physical containment sets up special mechanical and behavioral barriers against the spread of propagable infectious organisms beyond the experimental area. Such barriers have been steadily improved during the years in which known infectious pathogens have been studied. Filtered ventilation, inactivation or careful disposal of wastes, controlled access of personnel, and other measures are combined to isolate the experimental area. Experience shows that such measures can very substantially reduce risk *but are not totally foolproof.* In the case of new recombinants that are conceivably highly dangerous and might propagate under natural conditions, physical means alone may not guarantee against occasional breaches of local containment. Again, the actual consequences of such breaches can only be speculated about until they are suitably tested. Clearly, experimental releases of potentially dangerous organisms into natural environments should not be carried out until the conditions for maximal safety can be more clearly defined. The current guidelines forbid such experiments.

Biological containment refers to the use for recombinant-DNA experiments of hosts designed to survive and propagate only under the special conditions provided in experimental laboratories. Theoretically, such organisms, even if they surmounted physical containment barriers, would be unable to survive, to propagate, or to be transmitted in noncontrolled natural environments and could therefore cause no damage. Organisms requiring nutrients not found in nature or requiring temperatures far removed from animal body temperatures are examples. Geneticists are able to design and "construct" such organisms, but there are concerns about the effectiveness of this kind of containment in natural habitats where new genetic combinations with "wild" organisms are theoretically possible. There is ample evidence, however, that with time and particularly in combination with physical containment biological measures can substantially contribute to risk reduction.

Finally, there are sound grounds to expect that shotgun experiments into *E. coli* will vary in degree of danger depending on the donor species being analyzed. With respect to human health, the more distant the relation to humans the donor species has the smaller the risk. To do shotgun experiments with human DNA in the present state of knowledge would evoke the greatest concern, since covert human pathogens or deleterious genes might be multiplied and accidentally released. Other primate species seem somewhat less risky, though still distinctly questionable, since about 90 percent of the DNA sequences among other primates are similar to those of humans. Nonprimate mammals seem safer, and other vertebrates still more so. Invertebrates and plants appear to be the least threatening of all in terms of human health. Even this kind of general expectation has uncertainties, however, since many organisms evolutionarily distant from humans produce toxins (for example, bee or snake venom) or share pathogenic parasites (mosquitos) with humans. Particular genes from these species might therefore be dangerous and might be amplified by the shotgun approach. Moreover, focusing on human health fails to consider comparable dangers to other species and, therefore, to ecological balance.

Given these considerations, shotgun experiments involving complex organisms, particularly primates, clearly involve considerable uncertainty and therefore require special caution. Accordingly, primate shotgun experiments were not authorized under the original NIH guidelines, pending the development of higher levels of biological containment than were then available. It has been suggested, in fact, that such experiments not be undertaken except at specially designated sites that would then act as controlled sources for the clones produced. This would prevent the repetition of whatever risk might have been inherent in the initial cloning of unknown segments of entire genomes.

How Safe Is *E. coli*?

Shotgun cloning of an entire genome illustrates the problems raised by a particular *kind* of experimental exploration. Comparable problems are associated with the choice of a host organism for the experiments. Biologists found out long ago that in pursuing a particular question the selection of the most advantageous organism can be the difference between success and failure. Mendel discovered his famous ratios partly through his fortunate choice of sweet peas. The role of chromosomes in heredity was elaborated thanks to the perspicacity of T. H. Morgan in selecting the fruitfly. A large step toward molecular genetics was taken when Beadle deliberately selected the breadmold to fit his experimental requirements. So it has been with the colon bacillus *E. coli*. It has performed brilliantly as the subject for genetic studies of molecular mechanisms, including the discovery of recombinant-DNA techniques. Suddenly, however, its continued use has been challenged — not because it no longer is able to answer experimental questions, but because of its "family background." Since *E. coli* is characteristically found in large numbers in the human digestive tract, the concern is that it can therefore too easily reinfect humans, carrying with it whatever deleterious recombinants it may contain. Why not shift to host organisms less likely to infect the human species?

How great is the actual risk of continuing to use *E. coli*? *E. coli* is a normal inhabitant of the digestive tract not only of humans but of other animals. Generally speaking it is not a pathogen, but certain strains have been clearly implicated in serious infant diarrhea and in *"la turista."* Genetic exchange occurs in nature both among different *E. coli* strains and with other bacterial species, involving not only the chromosome but plasmids and viruses (phages). The details of these exchanges in natural environments are not well understood. Clearly they are important to assessing the consequence of possible escape of recombinants into natural ecosystems.

It is important to stress that the much-studied laboratory strain of *E. coli* (K−12) is markedly different from its "wild" progenitors. It is much less infective to humans, even on direct ingestion in considerable quantity, and produces no known harm. As noted earlier, it has been strongly argued by some in early stages of policy discussion that strain K−12 already provides significant biological containment by virtue of its considerable alterations during laboratory "domestication." Nonetheless, others have pointed out that more positive containment could probably be achieved by developing strains specifically dependent upon laboratory conditions. Again, the use of strains that do not grow at human body temperature or that depend upon exotic nutrients not found in nature would considerably reduce the risk of extralaboratory survival or transmission. This is the rationale for producing EK−2 biological containment as specified in the NIH guidelines.

There are at least equal difficulties with developing alternative species to *E. coli* as recombinant hosts. At the outset, all the experience and knowledge of *E. coli* accumulated over several decades would have to be abandoned. Building such a knowledge base for another organism might not take as long as with *E. coli* but nonetheless would require significant delay, cost, and duplication of effort. Moreover, it is not at all certain that any other species would be as advantageous experimentally as *E. coli* (though several possibilities have been suggested) or that the kind of objections raised against laboratory-weakened ("crippled") *E. coli* would not also be applicable to whatever other organism might be selected.

Is There Risk in Mixing
Two Genetic Domains?

We have already referred to Robert Sinsheimer's concern that the use of any species of bacteria in studies of complex organisms leads to a third kind of risk: that which may be involved in mixing DNA from two fundamentally different kinds of organisms. The living world falls into two broad categories, the procaryotes and the eucaryotes. The procaryotes include the bacteria and blue-green algae — relatively simple, basically single-celled, and without an enclosed nucleus. The paleontological record shows that procaryotes evolved before eucaryotes. These simple organisms probably made up all of life on earth for some 2 billion years. The first signs of eucaryotes appear in the palaeontological record at something over a billion years ago. At first these organisms were also relatively simple and basically single-celled. They had, however, an enclosed nucleus, chromosomes, and a mechanism for partitioning identical chromosomes to daughter cells. This latter

process by itself must have represented a very great modification and complication of the genome and its mechanisms of expression. It appears also to have been a prerequisite to the varied complexity we see around us today. For all complex organisms, both plants and animals, are eucaryotes, and most of the major kinds made their appearance more than 600 million years ago in what looks to be, on a paleontological time scale, a sudden efflorescence.

The argument is that evolutionarily procaryotes and eucaryotes have been separated for a long time, that their mechanisms of genetic expression are presumably very different, and that it may be part of a fundamentally evolutionary strategy that the two never intermix genetically. Hence it is argued that it may be insidiously destructive in biological terms to produce genetic combinations that nature has carefully forbidden through ancient barriers to interbreeding. To introduce into procaryotes (for example, *E. coli*) genetic control mechanisms that were specifically evolved by eucaryotes (e.g., humans)—conceivably in their early stages specifically to compete successfully against the procaryotes—may be very risky.

The argument is speculative but illustrates the deep doubts that unprecedented exploration can raise. It is reminiscent of the warning recently given by Sir Martin Ryle, astronomer royal of Great Britain and Nobel laureate in physics. He cautioned that radio signals emitted deliberately from the earth to establish contact with extraterrestial intelligence might be picked up by hostile beings who might decide to invade us. He was challenging the kind of equally speculative thinking illustrated by NASA's Project Cyclops, whose proponents have for nearly a decade been seeking increased financial support for efforts to listen for radio evidence of intelligent life elsewhere. Their theory is that we will have much to learn by communicating with what could be expected to be more advanced civilizations. These contrasting views are both based on scanty evidence; it makes the existence of extraterrestrial life plausible but is entirely inadequate in showing that such life is intelligent. To proceed beyond, with assumptions as to the friendliness or hostility of such intelligent life, truly enters the realm of science fiction.

Sinsheimer's warning of danger in eucaryote-procaryote recombination is equally speculative and equally worthy of consideration. In unprecedented exploration unprecedented consequences cannot be brushed off as simply alarmist. They cannot, however, be allowed to paralyze all action. Unlike theories about possible extraterrestrial life, procaryote-eucaryote recombination can be carefully explored here on earth—albeit with some risk. In fact, an almost equally speculative counterargument is that genetic exchange has probably already been tested many times in the complex ecologic interactions of procaryotes and eucaryotes. There is even a plausible theory that certain organelles of higher cells are really permanently incorporated procaryotes. In such speculative impasses it is necessary to formulate the difficulty more precisely and to design experiments that can be carried out under safely controlled conditions. A general hypothesis of conceivable but poorly specified infinite risk can only inhibit all action. Specific hypotheses of particular risks, however, can be significant incentives to creative thinking. In the process of investigating such questions we may learn much about both eucaryote and procaryote genetic control systems and the compatibilities and incompatibilities between them. Fortunately, the experimental analysis has already begun.

Are There Safer Alternatives?

The fourth kind of issue stems directly from those recounted in the preceding sections. If real though incalculable risks are inherent in insertion of recombinant DNA from higher organisms into living bacterial hosts, can the presumed advantages of this form of research be gained by alternative and safer means? Once again we face an uncertainty resulting from incomplete knowledge. There is, however, no question that alternatives to consider do exist. The hectic pace of advancing molecular genetics has opened many new doors. Recombinant-DNA technique is among the most attractive of options, and no other offers the same range or set of potentials. But in approaching particular questions other avenues certainly can be conceived and in some instances already are being explored.

For example, it is possible that with considerably less convenience and efficiency the mammalian genome can be explored in cell-free systems. DNA segments, in the presence of suitable enzymes, precursors, and energy sources, are replicated. Complementary messenger RNA can also be transcribed from them. When functional messenger RNA is isolated from a source organism it can, under suitable conditions, be used for reverse transcription of the appropriate genetic DNA sequence. It can also be combined with protein-synthesizing units (*ribosomes*), which it will instruct to make its corresponding protein. In fact, cell-free systems combining DNA and RNA effects have been designed that will begin with a specific DNA, synthesize its messenger RNA, and go on to produce the protein product.

Such cell-free systems carry out essential activities of living cells. However, they fall far short of actual living organisms. Therefore, they do not carry the risk of liberating independent and conceivably hazardous organisms into the environment. At the same time they do not afford the advantages of living bacteria in nature — their vigor and high efficiency — whose precision has long been honed by natural selection. It is just this natural adaptive efficiency that is weakened in laboratory strains to achieve safe biological containment of *E. coli*. In a sense, therefore, cell-free systems and development of genetically crippled *E. coli* strains converge toward a midpoint that may optimize both safety and efficiency. Patient research in this direction is urged by some as an alternative to the more direct investigation of recombinant behavior in *E. coli* using such techniques as shotgun cloning.

Are There No-Trespassing Zones?

Finally, we come to what is perhaps the ultimate issue in all exploration of the unknown. The four issues already discussed relate to events along the exploratory path — what the risks may be, how we might be able to respond, and what alternatives are open to us. The remaining issue has to do with what exploration may reveal, and whether there are unknowns that we should leave undispelled. Are there things to know that may fundamentally violate our own best interest? Are there ultimate mysteries better left undisclosed to a fallible humanity? *Ought* we to seek to breach these mysteries? This question is not only more difficult to answer, it is an entirely different *kind* of question.

Those who raise this question emphasize that the materials of heredity are the central core of all life, including that of the human species. Access to these materials provides the ability to intervene in processes that from the beginning of life on earth have been regulated by natural forces or the presumed divine hand. Proponents argue that human beings are not prepared for this access, that if they achieve it they may disturb essential relations and balances either unwittingly or malevolently, to gain private ends or to achieve irrational objectives. However phrased, this warning of danger goes beyond health hazard. It depicts capabilities that should lie in a nonhuman sphere and ultimate disasters comparable to a fall from divine grace.

Values clearly are involved in this issue and the answers therefore cannot be found in science alone. It is worth noting, however, what the *traditional* scientific attitude toward the problem has been, because molecular geneticists have faithfully adhered to it in their successful exploration of heredity. The method and habit of thought of science does not, in principle, accept a permanent *terra incognita,* no matter how fearfully such an image might be conceived. Cautious, yes — forbidden, no! The argument goes as follows. Science deals only with experiences that can be shared in common and that can be assessed by common criteria. We cannot commonly share the unknown; we can only compare our individual speculations, anxieties, and hopes about it. When we focus on a given area of the unknown, that area becomes significant. Provisionally, we seek to define this area through various hypotheses as to its nature. Science then moves to develop commonly acceptable evidence to test the alternative hypotheses. In the view of scientists the only true *terra incognita* consists of what we have not yet learned to recognize and conceive. No *recognized* area of the unknown can be barred from advancing knowledge.

The scientific tradition, further, does not regard knowledge itself as either good or evil; to acquire it is good and only its uses may be evil. Phenomena cannot be made more dangerous by our understanding of them; in fact, knowledge may be used to make them less so. That being the case, *lack* of knowledge is at least as dangerous as possession of it. To be sure, knowledge may be used to create harm or danger, but this manipulability only illustrates the neutrality of knowledge itself. It is the use, not the knowledge, that is the proper subject for value judgments. Thus, the search for knowledge cannot logically be restricted because of our potential for using it in harmful ways. If we always make the right decisions (unlikely though this may be) the uses can all be beneficial.

This is the traditional scientific perspective, threaded through the long history of scientific exploration. However evaluated in today's rapidly changing and uncertain world, the thrust of exploration is always toward the next unknown. Risks, in these terms, are only surmountable obstacles. They can impose delays, increase costs, and require greater ingenuity, but they can never block indefinitely the continued expansion of knowledge.

Molecular genetics has defined the gene, discovered the structure of DNA, and learned to recombine genetic determinants outside of cells. Left to its own devices, the exploratory thrust will not now stop short of understanding the human genome. In the process, an unknown will be dispelled and new capabilities for human intervention in, and control of, our destiny will be created. This is the attitude and thrust of science, now clearly a major contributor to the total human perspective. We shall turn later to the problems of appropriately blending this powerful voice into the full human chorus.

CHAPTER FOUR

Making
Better
Mousetraps

Recombinant-DNA techniques open a path to greater genetic knowledge. The simple search for understanding sometimes is said to be motivated solely by human "curiosity." Curiosity alone, however, could only in recent times be regarded as the chief motivator for the growth of knowledge. In the early history of our culture, curiosity could hardly justify time spent on the generating of knowledge for its own sake. The capacity to learn by investing time and effort in organizing and analyzing experience could only have been valued because it gave competitive advantage for *survival* to primitive humans. Learning from experience may have been "fun," but in the tough terms of evolutionary progression it was more likely that "getting things done" was what counted. Only the affluent societies of relatively recent times could consider curiosity as sufficient justification for serious work.

Therefore, the claim that knowledge has value for itself is legitimate, but it is not a full statement of the role of knowledge cultivation in human society. Today the *utility* of knowledge again is increasingly emphasized. This is because generating knowledge has become ever more costly, is funded more and more from the public treasury, and in many instances is suspected of leading to undesirable as well as beneficial consequences. Accordingly, the doctrine of knowledge for its own sake is challenged in many quarters as arrogant, narrow intellectualism. Demand is rising for justification of new knowledge in terms of social benefit. Knowledge is even regarded as a commodity whose investment costs must be assessed in terms of social return. Given its high cost, new knowledge can no longer be justified as attractive ornamentation; it must "pay off." Better mousetraps used to be expected from inventors and entrepreneurs. Today they are expected from the efforts of scientists, engineers, and even scholars.

Are recombinant-DNA techniques likely to produce better mousetraps? If so, how would the potential benefits of these products measure up to the risks and other costs of achieving them? Ideally, by using this utilitarian approach we would find a clear gain on weighing the aggregate value of the improvements against the aggregate costs, including the cost of insurance against risk. But this is not an easy calculation to make, even for simple mousetraps. And with respect to the possible products of recombinant-DNA research it is an exceedingly difficult calculation to carry out convincingly. Particularly in these early stages, there is as much conjecture in calculation of practical benefits as risks. Nonetheless, the utilitarian point of view requires some sort of estimate, for if the risk were determined to be very high and the hoped-for benefits negligible at best, social support of recombinant-DNA research might be judged to be unwarranted.

Projected benefits for recombinant-DNA research are, however, impressive. The danger lies in our taking favorable projections too literally, which could lead to disillusionment if they proved more optimistic or difficult to realize than expected. Controlled breeding that exploits natural gene recombination has yielded significant benefits for a long time in agricultural crops and domesticated animals. Recombinant-DNA technology holds promise for even more effective results over a greater species range. If a fraction of the promise is realized the benefits will be spectacular.

Bacterifacture of Human Proteins

The practical application of the technology most often discussed and likely to be earliest tested and achieved is the controlled production of human proteins by bacteria. This use would flow directly from some of the procedures described in the preceding chapter. The first steps toward application are actually already being taken.

The need for human protein arises whenever a deficiency of protein produced by the body causes human disease. Many examples might be cited, ranging from antibody formation to the production of natural painkillers. Hormones are a good example because they are essential blood-borne regulators of body function, and a number of them are proteins or smaller chains of amino acids. Pituitary dwarfism is a disease that results from a deficiency of the growth hormone produced by the pituitary gland, and diabetes involves a deficiency of insulin from the pancreas. Deficiencies or abnormalities of essential proteins of the blood also account for human diseases — for example, clotting factors in hemophilia or hemoglobin in sickle cell anemia.

The composition of proteins is highly specific, and this characteristic poses a special problem in providing replacements as a means of human disease therapy. As has been pointed out, the specific sequence of subunits in a protein reflects the genetic sequence of DNA. These sequences are important to the protein's shape and configuration, and these features in turn, determine its function. But shape and structure also determine the protein's capability to stimulate the recipient's immune mechanisms if it is transferred from one individual or species to another. This latter capacity could be both an advantage and an obstacle in the production and use of proteins. An advantage, for example, would be that proteins could be produced that form the outer coat of viruses and that make possible the entry of the viruses into cells. Such proteins could be used as cheap and safe vaccines to immunize against viral diseases otherwise difficult to control. The obstacle lies in the fact that hormonal protein from sources other than humans can act as antigens and thus be rendered ineffective by the immune system. At the present time the chief sources of insulin to treat human diabetics are pigs and cattle, because their insulin differs from human insulin in only a single amino acid. Insulin from other species is not only less likely to function as effectively but also may produce complicating immune reactions if given to humans. Problems of future supply, of pharmacologic effectiveness, and of immune compatibility would all be simplified if large amounts of human proteins could be produced rather than having to be extracted from human tissues themselves.

Research in this direction is proceeding at this time. Animal proteins are being produced as models in developing the methods. For example, rabbit globin is a protein that has been extensively studied, and its structure is well understood. Recently, recombinant-DNA techniques have been developed that are on the path to production of rabbit globin by bacteria. The first required step is to insert the DNA sequence, or gene, for rabbit globin into bacteria where it may replicate and function. To do this one must first identify and isolate the necessary gene. This is not easy in rabbits (or humans), since a particular gene is one of thousands within the DNA molecule and neither its location nor sequence may be known.

One can, however, take advantage of the complementary relationships we have described between genic DNA, its messenger RNA, and the protein eventually produced. The production of large amounts of the protein is the objective, but the corresponding DNA is in minute quantity and by ordinary chemistry is difficult to identify and separate from all the rest. However, under appropriate physiological conditions large amounts of the messenger RNA for globin are produced in the blood-forming tissues of adult rabbits. The specific messenger RNA for globin can be extracted from such tissues in significant quantity. This substance is, of course, complementary in structure to the genic DNA from which it was derived.

The transcription of complementary messenger-RNA sequences *from* genic DNA sequences was an early fundamental discovery in molecular genetics. The technique depends upon the action of special enzyme proteins, the transcriptases. However, not until attention focused on viruses composed of RNA was a reverse process discovered. When RNA viruses infect cells their replication requires that they first be transcribed in reverse to DNA. To do this the virus makes an enzymatic protein called a reverse transcriptase. Such enzymes promote formation of DNA that is complementary to a given messenger RNA (see Figure 4-1). Reverse transcriptase can be isolated from virus-infected cells. The enzyme works not only on virus RNA but on messenger RNAs produced by higher cells in normal genetic function. DNA complementary to messenger RNA can function in bacteria as though it were the gene required to produce a given protein.

If nucleotides, messenger RNA for rabbit globin, and reverse transcriptase are combined in the laboratory, DNA complementary to the messenger RNA will be produced. Thus, the genetic information necessary for the production of rabbit globin can be obtained by a much more direct and effective process than extracting it from the total DNA of the rabbit.* The amount of genetic DNA obtained by these methods is not great, but it is sufficient to be a source for recombination with *E. coli* plasmid DNA. The resulting recombinant plasmid can then be reinserted into bacteria to replicate. In this way the original amount of rabbit globin genetic material can be enormously amplified. The same procedure has been shown to work for the genetic DNA for rat insulin. What has not yet been shown is that these mammalian genes in bacteria will actually direct the production of the mammalian proteins.

Simpler mammalian genetic information, however, can now definitely be said to produce mammalian products in bacteria. Somatostatin, a chain of only 14 amino acids, is an important hormone in the communication between the human brain and the pituitary gland. It is, therefore, a significant pharmacologic agent. The sequence of its particular amino acids was determined after the substance was laboriously isolated from thousands of sheep brains. By determining the bacterial codon for each amino acid, researchers could chemically synthesize a DNA containing the appropriate nucleotide sequence coding for somatostatin. On insertion via a plasmid into *E. coli*, the bacteria produced somatostatin in large quantities, and on isolation this material

---

*Recent information indicates that the actual extracted gene would contain sequences of unknown function that are not determinants of the amino acid sequence of globin. Thus, the "native" mammalian gene can be replicated in *E. coli* but may not be translatable into protein.

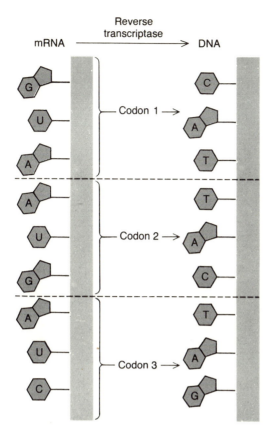

FIGURE **4-1**

Reverse Transcription of Messenger RNA.
Most often, a DNA sequence is transcribed to a
complementary RNA sequence with the help of an
enzyme known as a "transcriptase." In special cases
reverse transcription occurs, from RNA to DNA.
The enzyme involved in this process is a very useful
tool because it can produce a replicable DNA
message from the messenger RNA for a given
protein. Since RNA cannot replicate, this becomes a
way to obtain large amounts of replicable information
to produce a desired protein.

was pharmacologically active in mammals. This was the first fully successful bacterifacture of natural mammalian agents using recombinant-DNA
techniques.

This success with a short amino acid chain fortifies the expectation
that a solution can be found for the production of full-length proteins. It
is well-known that in bacteria genetic transcription to yield a messenger
RNA requires more DNA than the complementary sequence of the gene
itself. As discussed earlier, each gene has neighboring nucleotide sequences that act as "control switches" (see Figure 3-2). The neighboring

sequences do not code for part of the protein but interact with other materials in the cell that act as signals to start and stop genic function. In bacteria some such control sequences can be transferred from their normal gene associations and reassociated in new ones. This was, in fact, done in the successful somatostatin case. Before mammalian genes are inserted into bacteria they should be attached to such known bacterial control sequences. A recombinant involving not only a gene but appropriate control elements might then not only replicate, but also be able to be "turned on" to make its complementary messenger RNA within the bacterial cell.

This process would not in itself assure protein production. The messenger RNA thus produced would also have to activate the bacterial mechanisms for protein production. This may present further obstacles when a large messenger RNA comes from an entirely foreign source. We may not be able to overcome such obstacles until we have more detailed knowledge of the mechanisms of protein production in both bacteria and higher organisms.

The scenario for producing at will proteins of higher organisms in bacteria therefore is well within theoretical formulation and has been realized in a simple case. One cannot predict how long it might take to accomplish more complex feats or whether unanticipated insurmountable obstacles will be encountered. However, should the general scenario be realized, it will certainly provide major benefits. Proteins are key molecules in all biological systems, both as enzymes that regulate the construction and breakdown of substance and as major structural elements that provide the shape and configuration of bodily form. They are so highly specific in properties that they are intimately linked to their subunit sequences, and these defy largescale production by currently available chemical methods. The bacterifacture of proteins would be an enormous boon not only to medicine but conceivably to production of food and other materials as well. Therefore, though the degree of possible success remains to be established, potential benefits are clearly great enough to motivate and justify very considerable effort to achieve them.

Manipulation of the Genetic Apparatus

The fact that bacterifacture by transferring genes from higher organisms to *E. coli* is a technology already on the way must be credited to the vast store of knowledge acquired in the past quarter-century about the relatively simple genetic system of *E. coli*. However, equivalent manipulation of the hereditary properties of higher organisms by recombinant-DNA techniques is a very different story. Some possibilities for this bear reasonable discussion, but they are neither assured by current knowledge nor, generally speaking, likely to be developed in the near future. The obstacles to be overcome involve not just practical details but theoretical questions as well. Recombinant techniques to modify plant and animal heredity in a major evolutionary way is still more science fiction than science. Yet it is true that much science fiction of the early part of this century has proven to be the science of the present.

The theoretical problem begins with a means to introduce extrinsic DNA into cells of higher organisms in a way that can lead to replication and expression of the inserted material. Approaches that are being considered are based on experience with *E. coli*. At the moment, there are no available plasmids of higher organisms to serve as vectors, though there is some evidence that plasmids may exist in higher cells. More available and better known potential vectors are plant and animal viruses. As with the phage viruses of bacteria, the genetic information carried by plant and animal viruses is expressed in invaded cells to produce new viral generations. In some instances the viral information is incorporated into the host genetic system and propagates along with it. These are desirable characteristics for a successful carrier of recombinant DNA. They make viruses a prime candidate for applying recombinant techniques to higher cells.

Animal and plant viruses, however, are largely known because of their unfavorable effects on their hosts—that is, the most studied ones are disease-inducing and are referred to as pathogens. Either nonpathogenic viruses must be identified (their existence is regarded as likely) or the pathogenicity of the known viruses must somehow be neutralized if they are to be used as carriers. The viruses particularly must not be pathogenic for humans or capable of spread in natural environments. Only two animal viruses are approved under the NIH guidelines as meeting these requirements at the present time. One, known as SV–40, was originally derived from monkey cells in laboratory culture and has been extensively studied because it produces cancer in animals or cancerlike changes in laboratory-cultured animal cells. The evidence is against it having similar effects in human beings, although it is known to invade and propagate in cultured human cells and to produce changes that make them look more like cancer cells.

The second virus available for experimentation under current regulations is known as polyoma virus. It was originally obtained from cultured mouse cells and was shown to induce tumors on inoculation into newborn mice. It also induces tumors in several other animal species but has not been shown to induce human tumors (no virus has been shown to do this although some involvement of viruses in human tumors has long been suspected).

Because both the SV–40 and polyoma viruses have been implicated in the origin of some cancers, a great deal of information about them has been accumulated. Moreover, application of recombinant techniques to these viruses may shed light on the mechanism by which they induce tumors as well as make them available as vectors to transmit recombinant genetic information. Studying them in this way, however, involves some degree of risk, and the NIH guidelines require moderate to high containment facilities for any effort to use them as vectors. Their successful use on laboratory-cultured cells has already been demonstrated. Current studies are directed toward a more accurate assessment of the risk involved in using them and toward finding ways to minimize it. It is worth recalling, as noted in Chapter One, that it was the proposed use of SV– 40 in recombinant-DNA experiments that produced initial concern among involved investigators and led to the present degree of regulation of the entire field of investigation.

Assuming that these or other viruses can be used as recombinant-DNA carriers into cells of higher organisms, there is much to be learned about the genetic system of higher organisms in artificial culture. The path leads toward an understanding of the organization and operation of the human genetic system. We saw that this problem is likely to yield to recombinant techniques using bacteria as hosts for replicating genes of higher organisms. If we follow this path, without ever attempting to manipulate the genes of *intact* organisms we are likely to find clues and possibly even a full understanding of how higher cells turn their genetic messages into actual hereditary properties.

Genetic Modification of Intact Organisms

Extending such studies to intact organisms, however, poses serious complications. In higher animals, such as humans, modifying the heredity of a particular cell does not necessarily modify the heredity of the offspring of the complex individual of which the cell is part. This is because higher animals, although they arise from the single fusion cell produced when a sperm fertilizes an egg, subsequently become multicellular. The initial fusion cell divides to produce millions of cells. Early in the individual's development a few cells are set aside to produce sperm or eggs. These *germ cells* do not participate in the main developmental process but move into the developing testis or ovary when these begin to form. There they remain relatively quiescent until sexual maturation occurs. The important consequence is that only genetic change in these *germ-line cells* can be transmitted to subsequent generations. Genetic change in any other cells will be manifested only in their *cellular* descendants within the individual involved and not in his or her offspring.

This distinction becomes a fundamental consideration in the potential uses of recombinant-DNA techniques in higher animals. The germ-line cells are not only relatively few in number but also are relatively inaccessible within the organism, and our knowledge of their properties is restricted. They may ultimately be reached by intervention techniques, but at the moment the approach likely to be most rewarding for recombinant-DNA experimentation is the use of laboratory-cultured cells. Germ-line cells have yet to be successfully cultured as proliferating populations, and we do not know whether this will prove possible. Meanwhile, the proliferating populations of other animal cells available in culture are the best candidates. The genetic changes that occur in these, however, will only be transmitted in *cell* generations. Theoretically, the changes can be transmitted to subsequent generations of intact individuals via nuclear transplantation or cell fusion in very early embryos, but the success of such procedures in achieving genetic change has yet to be demonstrated.

Another consideration limits the range of practical application of recombinant-DNA techniques to higher animals, including humans. Most of what we know of how genes are expressed comes from studies of particular proteins and a particular gene responsible for them. Such favorable cases are relatively easy to find and study in bacteria, and they also are known from studies of higher organisms. For example, a number of human diseases have been attributed to single gene defects. In sickle cell anemia it has been possible to correlate the gene defect directly with a single altered component of the protein it produces. However, the complexity of development of higher organisms makes it more usual that inherited characteristics depend upon many proteins and the expression of many genes.

Therefore, change of more than one gene may change a given character and a single gene may affect many characters. The fundamental direct relationship between gene and *protein* undoubtedly exists in higher organisms, but the relationship between gene and genetic *character* (phenotype) is less direct and more complicated. In human beings, physical attractiveness, size, personality, intelligence, and many aspects of disease susceptibility, are "polygenic" characteristics. Many genes relevant to a given character are scattered on the 46 human chromosomes. Inheritance of such polygenic traits is not fully predictable in human pedigrees, and such traits are unlikely to be fully controllable by insertion of small specific fractions of the total genetic information responsible for a given individual. Thus, the concern that recombinant-DNA techniques may soon allow the artificial production of super-races ("alphas") or infra-races ("omegas") in the human population is both an oversimplification of human heredity and an exaggeration of current knowledge. An effective approach to modifying such human characteristics, even if this were judged desirable, is not within practical biological range today. Recombinant-DNA techniques may be an early step in that direction, but this step no more leads inevitably to full realization than the first step on the moon inevitably meant that human beings will travel to and colonize distant galaxies. Much more will have to be known and many more opportunities for decision will present themselves before recombinant-DNA techniques can possibly be used to alter broadly the nature of humanity.

However, some more limited applications of the technique to humans are possible. Repairing the effects of defective genes in individual people should be less difficult to accomplish than controlling human evolution, as well as more easily accepted as a benefit rather than a threat. Genetic intervention in individuals need not entail transmission to subsequent generations. Theoretically, using this technique to cure disease would be similar to employing any other therapeutic device. Sickle cell anemia is a case in point. This disease involves abnormality of red blood cells due to a genetic defect in the protein globin. Symptoms appear fully only in individuals whose red-blood-forming cells contain two defective globin genes, one from each parent. The parents have one defective gene and one normal one in their red blood cells, and they only occasionally show clinical symptoms though they can be detected as "carriers" by suitable tests. Clearly, one normal gene for hemoglobin per cell is sufficient to keep an individual reasonably healthy, but offspring who get two defective genes from their parents are severely afflicted. The disease is serious enough that afflicted individuals may not survive their early twenties.

Suppose normal human globin genes were produced in quantity in bacteria using recombinant-DNA techniques as described for rabbit globin. Suppose these genes were then combined with a virus able to carry the genes into human cultured cells. The possibility would then exist of culturing immature red blood cells from the bone marrow of a sickle cell patient and of infecting them with the harmless virus containing normal globin genes. These infected cells might then be reinoculated into the patient's bone marrow. Such inoculation of bone marrow cells under suitable circumstances has led to colonization and repopulation of a patient's bone marrow with descendants of the injected cells. Thus, this procedure — at the moment only a matter of speculation — might yield effective therapy by recombinant-DNA techniques to allow the individual patient to produce normal mature red cells from his or her own immature cells. Such genetic repair would be nontransmissible to offspring but could give the patient an improved and possibly normal life. With variations of technical detail this technique might be applicable to other gene-defect diseases, particularly where essential enzymes are abnormal because their responsible genes are defective.

Techniques of this kind might be applied even more easily to plants, and with greater chance of having a desired effect not only on individuals but on descendants. Transmission to descendants is easier in plants because they have greater capability than higher animals for nonsexual propagation. As an example, plants propagate by sending out "runners," each of which can produce a new plant that is genetically identical with the original. Another example is propagation of plants by "cuttings," that is, starting new plants from portions of the stem of older ones. In addition, when appropriate plant parts are cultured in the laboratory even a single isolated cell can sometimes give rise to a whole new plant.

Further, plants are infectible by viruses just as are animals. A well-known plant virus produces "mosaic disease" in tobacco plants. Such viruses are potential vectors for recombinant DNA. To be used for these purposes, of course, these viruses would have to be modified so as to be made harmless to their hosts, and they would have to introduce the gene into a site where it could express itself. If these requirements

could be met — and there are significant but not theoretically insuperable obstacles to be overcome — transfer and recombination of plant genes might produce significantly improved yield or disease resistance in crop plants.

Among the possibilities widely discussed is that of conferring the capability for nitrogen fixation on crop plants that now require external addition of expensive fertilizers. Nitrogen-containing compounds are essential constituents of all living organisms, and free nitrogen is the most abundant component of the earth's atmosphere. Nonetheless, only some of the simplest and most ancient living things — bacteria and blue-green algae — are able directly to utilize atmospheric nitrogen by fixing it through reduction to ammonia. Once fixed, nitrogen becomes available through food chains, built into the many other compounds made and utilized by all living organisms. Thus, the supply of fixed nitrogen is one limit on the production of biomass in any given ecosystem, whether natural or artificial.

Agriculture can be thought of as the maintenance of artificial ecosystems that yield food for human beings. Intensive agriculture with essentially pure stands of one species usually exceeds the natural availability of fixed nitrogen and can only be practiced through supplementation of the fixed nitrogen supply. A number of practical means of supplementation have been devised by practical and scientific agriculture. In the most intensive agriculture, as carried out in technically advanced countries, the dominant procedure is addition of chemical, fixed-nitrogen fertilizers. These may come from natural deposits (which are rapidly declining) or from chemical industry. Both sources are increasingly expensive in dollars and chemical manufacture is heavily dependent on dwindling energy sources. Accordingly, there is growing interest in increasing biological capability for nitrogen fixation.

The problem is not simple and is being investigated in a number of ways. Recombinant-DNA techniques may be applicable because the capability to fix nitrogen is genetic. It is known that organisms that can carry out nitrogen fixation contain genes for appropriate enzymes. The bacterium *Klebsiella pneumoniae* is such an organism. It produces genetically deficient mutants that cannot fix nitrogen because they have abnormal enzyme proteins. Such strains can be used to obtain information on the genetic basis for nitrogen fixation. If, for example, plasmids of *E. coli* that contain DNA from normal *K. pneumoniae* are inserted into defective strains of *K. pneumoniae,* the mutant strains recover their capability to fix nitrogen. Such studies show that at least eight genes are involved in nitrogen fixation, three that work together to form a single necessary enzyme and five whose products cooperate with the enzyme or act as control factors for its production.

Judging from this information, it appears possible that the capability to fix nitrogen may be improved in bacteria normally having it or conferred on other bacteria (or blue-green algae) that do not have it.

Turning such information into a useful agricultural tool, however, has complications. Certain higher plants, for example the legumes, use bacterial capability to fix nitrogen by serving as hosts to the bacteria in special root nodules. In these circumstances nitrogen fixed by the bacteria is transferred to the higher plant in a coordinated and efficient way. However, bacteria that are free-living nitrogen fixers are less efficient in transferring nitrogen to associated plants. They may even become disadvantageous by giving off excessive ammonia to the soil and rendering it excessively alkaline. What is needed, therefore, is either the creation of new associations between nitrogen-fixing bacteria and higher plants, comparable to that involving legumes, or the introduction of nitrogen-fixing genes directly into higher plant cells. Neither of these routes will be easy and both can conceivably be accomplished by other techniques than recombinant DNA.

Two major uncertainties color the whole subject of improved nitrogen fixation by whatever method. First, nitrogen fixation requires a very heavy energy investment on the part of the organism carrying it out. The energy required is so great that the capability to fix nitrogen conferred on crop plants might result in an unacceptable reduction of crop yield. Second, a massive increase of released ammonia into soils due to increased nitrogen fixation for improved agriculture might have unfortunate secondary consequences on the oceans and atmosphere. Ammonia is not only toxic in itself but also tends to reduce atmospheric carbon dioxide levels, which in turn tends to reduce average world temperatures as well as ozone levels in the upper atmosphere. Reduction of average world temperatures by even one degree can counteract any favorable agricultural effect of improved capacity for nitrogen fixation. Reduction of ozone levels increases ultraviolet flux at ground level and would be expected to increase the incidence of skin cancer.

These facts add up to an interesting but still uncertain prospect with respect to nitrogen fixation, with recombinant-DNA techniques playing a possibly useful but not necessarily a crucial role. This conclusion regarding nitrogen fixation is not unlike the present overall conclusion for recombinant DNA as a source of better mousetraps. We have a new excellent candidate for intervention in genetic processes. It almost certainly will convey some benefits and it should not be lost as a possible tool. Its actual performance, however, can only be judged as knowledge moves ahead and priorities are assigned to our needs. Each potential new use should be assessed for consequences before a decision is made, and early stages of development should be monitored closely. Knowledge itself can never guarantee "progress" or provide absolute certainty. As the body of knowledge grows, however, it does provide new opportunities to test imagined futures. Which opportunities we accept should not be left to those who generate knowledge alone, but should be the product of decisions appropriately made within the body politic. We turn next to the reactions of the body politic to recombinant DNA.

CHAPTER FIVE

DNA Meets the Body Politic

Across the Charles River from Boston—the cradle of liberty famed for rebellion against both tea taxes and enforced bussing—lies Cambridge, home of Harvard, the Massachusetts Institute of Technology, and Mayor Alfred E. Vellucci. Cambridge is an industrial and academic city of about 100,000 inhabitants. It teems with blue-collar workers but also is rich in intellectual and scientific talent. Frequent tensions occur because of clashing ideologies among academic dons and between town and gown. Mayor Vellucci is a "townie" and cultivates the image of a man of the people. One of his suggestions for solving a Cambridge parking problem was to cut down the trees and pave the Harvard Yard. It didn't happen, but Harvard cars became fewer at the curbs of Cambridge streets.

Controversy has been no novelty in Cambridge in recent years. Student unrest in the sixties and sociobiology and human genetics in the seventies have focused attention on Cambridge as a center of intellectual ferment. In the case of recombinant DNA, issues centered around a seemingly routine academic split: Should an old biology building on the Harvard campus be modified to provide a P–3 containment facility?

The proposal to do so emanated from molecular geneticists. Objections came from other biologists, long at war among themselves and with their molecular biological conferees over changing emphases of biological investigation and social outlook. The debate over the P–3 facility spilled out of department conclave into a campus forum sponsored by the research policy committee of the university. In a parade of impassioned views the potential benefits and hazards were presented. Most apparent, however, was the absence of consensus among knowledgeable people. Following the campus forum the research policy committee recommended that the building be modified as proposed. A significant factor, it would eventually turn out, was the presence at the forum of a Cambridge councilwoman who rose during the proceedings to make herself known and to announce that the city council would be following the matter closely.

Indeed, Mayor Velucci speedily mounted the battlements. He had read a report of the campus forum in the Boston *Phoenix,* an "alternative" newspaper in the local community. He had also been visited by Nobelist and Professor George Wald of Harvard and his wife, Professor Ruth Hubbard. Both were outspokenly opposed to the proposed recombinant facility. Fortified by "scientists of his own," the mayor performed his civic duty with alacrity. "We want to be damned sure the people of Cambridge won't be affected by anything that would crawl out of that laboratory," he said. "It is my responsibility to investigate the danger of infections to humans. They may come up with a disease that can't be cured—even a monster. Is this the answer to Dr. Frankenstein's dream?"* The subject of recombinant DNA was put on the Cambridge City Council agenda for informational discussion on the evening of June 23, 1976, coincidentally the very day the NIH guidelines were formally issued.

The Political Forces at Cambridge

Other political forces fed the fire of the DNA controversy in Cambridge. One was an organization, consisting primarily of younger scientists and technical personnel, known as Science for the People. It had come into being in the Boston area several years before to provide a radical challenge to "establishment science." Its members believed that the connections of science with the industrial-governmental complex were carrying it in directions dangerous to the interests of the general public. The organization emphatically called for a more people-oriented perspective and took initiative to curb what it regarded as antisocial research directions. The organization's Genetics and Social Policy Group, for example, was in combat with "recent attempts to trace societal problems to the genes of individuals rather than to inequities in society itself."

*New York *Times,* June 17, 1976.

The group had successfully attacked research at Harvard on a possible linkage of chromosomal abnormalities with aberrant social behavior. It had also given attention to the DNA controversy from its inception. In addition to publicizing grave concerns about the possible health hazards of recombinant-DNA research, the group was attempting to organize involved nonprofessional personnel into "safety committees" intended to "confront the dangers to health that DNA-recombinant and other laboratory work presents." Its critique of the NIH guidelines in June 1976 — just as the Cambridge controversy erupted — stated, "In the name of improving human health, newer and more potent threats to human health are being developed. It is unclear that these genetic technologies have been developed in response to national needs or whether they are simply the interests of professional scientists who make their living with such developments."*

Also feeding the Cambridge controversy were environmental public interest groups. Francine Simring, on behalf of the Committee for Genetics of Friends of the Earth, wrote to Mayor Vellucci on June 16 calling for a moratorium on recombinant-DNA research. Citing a "growing controversy," she asserted that it stemmed not only from health concerns but from "the moral and ethical questions that arise if scientists tamper with the heredity and evolution of their fellow-humans." She urged the mayor to take every precaution "to ensure that a small group of scientists does not put its own interest before that of our citizens."

Not to be overlooked in assessing the political background of the Cambridge episode is the strong connection seen by many between the DNA controversy and the long struggle over the practical consequences of nuclear fission and fusion. Early in June, during an MIT symposium sponsored by Miles Laboratories, graduate student Frances Warshaw had called for a halt to recombinant-DNA research. She pointed to a correspondence she saw between genetic engineering and the atomic bomb, a relationship about which she felt keenly because both of her parents had participated in developing the bomb. She asserted that "nuclear energy has given us fear and horror and not many clear benefits." Similarly, L. Douglas DeNike, a Los Angeles psychologist active in emphasizing the need for security against possible terrorist uses of nuclear devices, warned Mayor Vellucci in a letter of June 17 that recombinant DNA would require "a level of security at least comparable to that of the nuclear industry."

The Cambridge City Council

The Cambridge Council Hall is not normally the site of epic events. The business of the council is generally mundane. Nonetheless, on the evening of June 23, 1976, the Council Chamber and adjoining lobbies were

*From a flier distributed by the group.

jam-packed, with a loudspeaker broadcasting the proceedings to those who could not get inside. Reports of what transpired in that meeting, which lasted until long after midnight, circulated not only across the nation but around the world. The assemblage in the Cambridge Council Hall was an unusual congregation of academics and concerned citizens. The presentations were not always brilliant and incisive but they were widely interpreted as signalling a changing relationship between science and the body politic.

Citizens, as town councillors, clearly were in the seats of decision and scientists of distinction but of diverse persuasion were before the bar. Recording the scene was the local and national press. It was a very different occasion from the one at Asilomar. DNA had clearly lost its veil of political chastity and stood fully exposed in the arena of political decision. In this arena uncertainty was not a source of excitement and challenge but of fear and anxiety. The timesaving technical jargon of scientific insiders had no place here; everyone had to use language accessible to citizens and politicians. If fundamental science was not before the bar of justice in Cambridge, it was nonetheless facing a kind of jury. The decision of the jury would have a heavy impact, whether or not the Cambridge City Council should place legal restrictions on two of the foremost academic centers in the world.

It is worth noting that though this informational civic hearing was initiated formally by the city council, the hearing would probably never have occurred had not controversy already existed on the substantial issues among Harvard biologists. The concerns of scientists had precipitated the Asilomar conference, and disagreements among scientists precipitated the Cambridge hearing. Public authority did not capriciously reach into the scientific domain; it was attracted by sounds of conflict and it found sharp division among the members of the scientific community.

Most of the councillors were encountering DNA for the first time at the hearing. In that initial discussion they heard explanations of its nature, they were told about the history of the controversy, they received formal reports on facts and procedures from Harvard officials, and they listened to widely divergent assessments as to both risks and benefits. In the wee hours of the morning they recessed the discussion, having agreed to come to some conclusion on the evening of July 7. Before them was a proposal by Mayor Velucci to ban recombinant-DNA research in Cambridge for two years, at least until all health hazard had been evaluated and brought under assured control. The following day the mayor modified his proposal to a voluntary three-month moratorium.

The next two weeks were filled with intensive activity on both sides of the issue. One councillor offered the opinion that citizens and councillors were having more contact with each other on this issue than they had on any other that had faced the city of Cambridge. Information booths were set up at two street fairs and proponents went directly to the people. Letters poured in from all over the country exhorting the council to act responsibly and decisively, both pro and con. There was a general feeling that as Cambridge went, so might go the country.

The evening of July 7 brought another jam-packed session. The two-week interval had been spent by councillors not only in assessing the mood of constituents but in seeking reasonable grounds for compromise. By a vote of five to three with one abstention, the council asked Harvard and MIT to observe a "good faith" moratorium of only three months while a study was conducted by a group to be known as the Cambridge Laboratory Experimentation Review Board. This board, consisting of lay persons and scientists uninvolved in the research, was to examine all issues in detail and to recommend appropriate further steps to the council. The compromise was accepted by MIT and Harvard and the board took up its burdens.

The Cambridge Review Board

After more than a hundred hours of meetings, the review board submitted its recommendations and findings to the city manager on January 5, 1977 (see Appendix IV). A second three-month moratorium had been provided to allow the board to complete its work. The board had been charged "to consider whether research on recombinant DNA which is proposed to be conducted at the P–3 level of containment in Cambridge may have any adverse effect on public health within the City. . . ." By deliberate intention the composition of the board did not include partisans with respect to the issues or persons expected to bring scientific expertise to the problem. Rather, the effort was "to create a committee of Cambridge citizens who could approach the subject in an unbiased manner and insure that the public safety is at all times the foremost consideration." In presenting its report, the board affirmed its belief, based upon its experience, that "a predominantly lay citizen group can face a technical scientific matter of general and deep public concern, educate itself appropriately to the task, and reach a fair decision."

The board concluded that recombinant-DNA research requiring P–3 containment could be carried out safely in Cambridge under appropriate conditions and safeguards. Required was strict adherence to the NIH guidelines with certain additions: (1) provision of a safety manual and a program of special training of involved personnel; (2) an appropriately broad-based membership composing the Institutional Biohazards Committee required by the NIH guidelines to oversee the research; (3) EK–2 biological containment for all experiments requiring P–3 physical containment; (4) appropriate testing and screening of all organisms used; (5) appropriate monitoring for infection and escape of experimental organisms; (6) establishment of a permanent Cambridge Biohazards Committee to monitor all recombinant-DNA research in the city. In addition to these local specifications the board called for federal legislation to strengthen and broaden the applicability of the NIH guidelines. The board's recommendations had been reached unanimously and they were endorsed, also unanimously, by the Cambridge City Council on February 7, 1977.

These tumultuous events in Cambridge clearly constituted a milestone in the interaction of recombinant DNA with the body politic. They brought sharply into public focus the major issue as it was then perceived, potential hazard to human health. The conclusions confirmed the basic approach of the NIH guidelines but indicated public preference for federal legislation covering all DNA research and requiring somewhat stricter containment. In effect, the Cambridge board said that what scientists had provided for control, using the NIH as an implement, was sound but that the public was not yet at ease and needed further assurance.

Other Political Foci

Discussions elsewhere, both before and during the Cambridge proceedings, were in substantial accord with the Cambridge conclusions. More than a year earlier the New York Academy of Sciences had held a conference highlighting social and ethical issues of recombinant-DNA research.* The serious questions raised had echoed in the debate over the guidelines (pp. 133 ff.) and recurred as a minor theme in later discussions. They did not, however, become a major focus of attention at Cambridge; the Cambridge Review Board excluded them specifically from consideration just as had been done at Asilomar. A similar outcome occurred at the University of Michigan, site of the first major-scale post-Asilomar review of recombinant-DNA issues. Concerns of a small number of faculty members led to the formation of a university committee with broad disciplinary representation, and the Michigan group undertook a local study early in 1975. Publicity was given to the study to elicit both on-campus and off-campus opinion; proponents, opponents, and critics had an opportunity to present their views. The resulting report, "Policy for the Molecular Genetics and Oncology Program," favored continuance of research up to the P–3 level under the NIH guidelines, regardless of the funding source. A university Biological Research Review Committee was called for to assure overall compliance, and a legal review found no special problems other than the ones of safety. The committee report was the subject of a formal critique by an opposition faculty group and there was debate both on and off campus. The University Board of Regents subsequently approved the committee report and the continuance of recombinant-DNA research up to the P–3 level of containment.

*Ann. N.Y. Acad. Sci., Vol. 265 (1976).

As a direct result of publicity attending the Cambridge City Council proceedings, Mayor Pete Wilson of San Diego directed his staff to look into plans for a P– 3 containment facility at the University of California, San Diego. Following discussions between the mayor's staff and university officials the mayor asked the Quality of Life Board of San Diego to conduct a study and report to the city council. A task force of the board that included scientists, a community member, and a representative of the County Board of Health heard testimony from a number of persons, locally and around the country, on both sides of the issue. At no time was intense public interest displayed, although a relatively small number of community members voiced strong opposition. The report of the task force resembled that of the Cambridge board, acknowledging the continuance of P– 3 level recombinant-DNA research and recommending continuance of the city task force to maintain community awareness. The board also sought several somewhat more stringent requirements than those of the NIH guidelines. The city council accepted the Board's report, and then the Board of Supervisors of the County of San Diego asked its Environmental Review Board to study the matter. A course of events comparable to those at Michigan, Cambridge, and San Diego also occurred in Princeton, New Jersey.

Meanwhile, activity at the state level was beginning in both New York and California. In New York the attorney general held hearings, bringing in experts of divergent opinions as well as public interest groups. In California, a joint hearing by two committees of the state assembly initiated by Assemblyman Charles Warren emphasized health as well as environmental concerns. The hearings led to the introduction of a regulatory bill in the California Assembly. In both California and New York the intention seemed, at least in part, to bring pressure for national legislative action. In Washington, Senator Kennedy had already held a hearing before his subcommittee on health, and the House Science and Technology Committee had initiated oversight hearings. Some uncertainty remained at the time, however, as to whether federal legislation was necessary and acceptable to the scientific community.

Public attention to the issue increased sharply early in 1977. Without any decisive new evidence of imminent hazard, media coverage mounted noticeably. Whether this was a delayed reaction to earlier publicity or the result of decisions by prominent environmentalist groups to give the issue higher priority is not clear. Nonetheless, an Academy Forum on recombinant-DNA research, arranged under the auspices of the National Academy of Sciences,* became the setting for public registry of political conflict and heightened urgency through activist demonstrations.

---

* See *Research with Recombinant DNA* (Washington, D.C.: National Academy of Sciences, 1977). The following quotations of speakers at the Academy Forum are drawn from this source.

A less likely setting for activist demonstrations would be hard to conceive. The National Academy building, directly across from the Lincoln monument in Washington, is constructed in the classic Greek mode. Usually it is the site of calm, reasoned deliberation and refined, almost courtly receptions. Under its central dome the slow, inexorable swing of a suspended pendulum reveals the diurnal rotation of the earth. The academy auditorium, carefully modulated in color and tone, conveys serenity and encourages contemplation.

Academy Forums are a relatively recent and still somewhat unsettling innovation in academy operations. Reflecting the increasing traffic between science and government, they represent an effort to develop open dialogue among science, government, and the public on policy matters of importance and even of controversy. The forum on recombinant-DNA research was arranged by a committee led largely by molecular geneticists. Their intention was to provide a "showcase" for varying scientific perspectives on the issue against the backdrop of political Washington. It is fair to say that the planning committee encountered some differences of emphasis as to how best to proceed. Furthermore, external pressures increased as planning moved forward.

The forum was scheduled for two and a half days on March 7–9, 1977. The opening evening session was highlighted and dominated by an orchestrated demonstration by activist groups who aggressively insisted on being heard. Word of possible problems had circulated in the corridors prior to the session, and the audience's expectation of something unusual was heightened on entry into the auditorium. The normally subdued lighting had been overwhelmed by glaring floodlights suitable for television broadcasting strategically placed around the room. Symbolically, every nook and cranny of the auditorium, every defect in architectural design or maintenance was naked and revealed in the pitiless light. Blinking and slightly confused, the speakers, panelists, and other invited participants found their way to seats in forward sections set aside for them, aware that a considerable audience of undefined character was ranged behind them.

The demonstration that followed was conducted by a relatively small group known as the "People's Business Commission," led by a spokesman named Jeremy Rifkin. This group had earlier condemned the forum program as "stacked," and demanded a broader presentation of views. They also accused the organizers of using tainted money from pharmaceutical companies to fund the forum. They were turned down on their demands for revision of the program but were conceded an opportunity to speak to the gathering. While Rifkin attacked his captive audience from the podium on the themes of Huxley's *Brave New World,* his supporters sang, chanted, and broke out banners of protest around the periphery, in ready range of the television cameras.

With the discovery of recombinant DNA scientists have unlocked the mystery of life itself. It is now only a matter of time — five years, fifteen years, twenty-five years, thirty years — until the biologists, some of whom are in this room, will be able literally, through recombinant-DNA research, to create new plants, new strains of animals, and even genetically alter the human being on this earth ... [T]his technology is sensational because it hits right to the basic emotional core of life itself. For three generations of Americans weaned on Huxley's *Brave New World* the long-range implications of experimentation in this field are ominous. ... Wait until the Protestants, the Jews and the Catholics, the Methodists, the Presbyterians and the Baptists all over America start to realize the long-range implications.

Rifkin's own ominous final warning was, "Let's open this conference up [to the people] or close it down!"

The demonstration within the meeting hall followed upon a press conference outside announcing the formation of the "Coalition for Responsible Genetic Research." Supporters included Nobelist George Wald of Harvard and Cambridge; Nobelist Sir MacFarlane Burnet of Australia; Lewis Mumford, writer and critic of technological impacts on society; environmentalist organizations such as Friends of the Earth, the Environmental Defense Fund, and the Natural Resources Defense Council; and the Cambridge-based group Science for the People. These individuals and groups, many of whom had been active in Cambridge, New York, and on other earlier battlegrounds, called anew for a total moratorium on recombinant-DNA research.

After the externally inspired fireworks of the first hours, the Academy Forum settled down to a more traditional course. Polarized dissension among scientists was continually evident. In an early session, Maxine Singer of the scientific staff of the National Institutes of Health sought to rebut the charge that scientists are so driven by the rewards of discovery that they pay little attention to public concerns. Scientists recognize their responsibility to the public that supports them, she insisted. "Dispute over the best way to exercise that responsibility must not be confused with the negation of it." While freedom of inquiry is a democratic right it is clearly unacceptable to cause harm in the name of research. But she warned that levels of anxiety are not necessarily directly related to levels of real risk.

Jonathan King of the Department of Biology of the Massachusetts Institute of Technology, representing Science for the People, quickly counterattacked. He characterized Singer's approach as a "technocratic coup" under the masquerade of scientific responsibility. He charged that the Asilomar conference was dominated by molecular geneticists, did not provide a sufficiently broad range of expertise, and consequently was biased in its perspective. He visualized "unbelievable conflicts of interest" among the participants there and subsequently during the development of the NIH guidelines. Erwin Chargaff, distinguished Professor Emeritus of Columbia University, joined the attack but pessimistically doubted that much could be done to stave off the dire threat of recombinant DNA. A Platonic community, he said, would certainly have regarded the genetic inheritance of mankind as its greatest treasure, to be protected at all costs from defilement. Instead, "we are sliding into an awful mess." Anyone claiming that immediate disaster is upon us is a charlatan, he said, but anyone who denies the possibility of disaster is an even greater charlatan. George Wald condemned the whole discussion of safety as misdirected. The question, he said, is not safety but whether to do the research at all. He characterized recombinant-DNA techniques as "manipulation and deformation of nature," unnecessary because alternative approaches are available. He described the whole argument for artificial recombination as a plea for mere convenience and speed, made without regard to potential corrosive consequences.

On the other hand, Daniel Nathans of Johns Hopkins University, Paul Berg of Stanford University, and others countered these dire prophecies by predicting both gains in fundamental knowledge and practical benefits. Better and cheaper pharmaceuticals, vaccines against viruses, enhanced crop production, genetic repair, cleansing of the environment were all mentioned, and in some instances elaborated, as conceivable paths of achievement. Thus, the by now familiar disagreements among scientists were recapitulated, disagreements that could only be resolved by more experiments and more facts than were currently available. Had this recapitulation been all that the Academy Forum produced it would have been no more than reiteration at the national level of what had occurred at local levels in Ann Arbor, Cambridge, and San Diego. But new elements were surfacing. These were still relatively minor themes in the cacophony of the Academy Forum, but they portended a new context to the issue in which not only recombinant DNA but knowledge itself might be in the trial box.

For example, Anthony Mazzocchi of the Oil, Chemical and Atomic Workers Union, spoke bluntly of workers who view the scientific debate in the context of their own experiences with technology and its consequences. Workers learn about occupational hazards from their own experience; scientists never warn them about risk even when they have data to suggest it exists. Workers are victims, never part of the decision-making process in any shape or form. Federal regulations and voluntary guidelines may be well-intentioned but they don't work to ensure safety in the workplace. When a new process has *possible* danger, that danger *always* materializes in the workplace and in workers' communities. Most occupational health and medical problems are not the business of science: they are economic and political. Such problems are detected by the primitive "body in the morgue" method. Given the experience of workers in the atomic and chemical industries, recombinant DNA should not be introduced industrially if *any* possible hazard exists. All outstanding questions must be resolved beyond the shadow of a doubt. Until that time, Mazzocchi concluded, continue the

debate!

From the philosophical side, Stephen Toulmin of the University of Chicago warned scientists not to underestimate the power of the public imagination to conceive "forbidden knowledge." He recalled the Faust legend and the travails of the alchemists of a few centuries ago when they were suspected of making "little men" out of lifeless materials. He noted that early atom-splitters had been looked at askance for tampering with nature's mysteries and that their successors were still struggling with often dubious control rationales. On the other hand, he noted, fire was not outlawed; rather, arson was defined as an antisocial act subject to legal sanction. How should the interests of the larger society be represented in decisions regarding the generation of knowledge? Toulmin asked. And what kind of institutional machinery is required to deal responsibly and fairly with those issues for which recombinant-DNA research "is only one early though difficult and contentious example?"

Tom Howard of the People's Business Commission reported on a small "town meeting" held during the workshop sessions on the second evening. He returned to the moral question: Should this research be done at all? "Who will decide the genetic fate of humanity?" he asked. "Who, here in this room, is qualified? . . . Are we ready to proceed, really post-human, beyond the human species?" He went on:

"This research that you have started is going to be carried to its ultimate conclusion. It is part of the technological society. . . . I want to let you know that this is just the first protest . . . we are just the little ruffling wind before the storm of public outrage . . . this is the most important social issue of the coming next decade . . . we are not going to go quietly. We have means at our command to resist the change in the human species. We will not go gentle into that brave new world, that new order of the ages that is being offered to us here.

Francine Simring, representing Friends of the Earth, an organization participating in legal action to halt federal support of recombinant-DNA research, concentrated on issues of safety, both human and environmental. In particular, she questioned the intentions of industry. She countered statements made by industry representatives at the forum that industry was intending to comply with the NIH guidelines by citing contrary statements from the record. She brought up the general problem of transport of hazardous materials, including recombinant DNA, suggesting that this problem was as important to the industrial sector as laboratory containment. She also called attention to the special problem presented by industry's insistence on confidentiality to protect its proprietary interest in newly discovered processes.

The critics of recombinant-DNA research, as represented at the forum, fell into two groups: dissident scientists, most but not all of whom were from the Cambridge-Boston area, and activist organizations, most of which had environmental interests. These two groups constituted the core of organized opposition on the national scene.

Their energetic efforts clearly had heavy impact on public opinion through the media and directly on Congress. It is of interest that comparable groups did not appear in other countries and public controversy nowhere else reached comparable levels.

Ranged against these groups at the forum were the molecular geneticists themselves and a smaller number of other scientists and nonscientists concerned about excessive restriction of fundamental investigation. Tracy Sonneborn, a distinguished nonmolecular geneticist from the University of Indiana, discussed not so much the substance of the forum, as the impression that the meeting itself made upon him. He emphasized first how complex and emotional the issue turns out to be. Second, he acknowledged that the public had a vital concern in many areas of scientific research and that it was therefore necessary to provide the public free access to all judgments and actions. Base must be touched, he said, with all of the interested groups or they may become implacable enemies. Third, he was depressed by the seeming inability of many scientists involved in the controversy to remain open-minded and objective and their willingness to substitute personal attacks on their opponents for reasoned discussion. He pointed out that this behavior could only inflame public anxiety and strengthen the trend toward antiintellectualism. Finally, he noted that the public choice is between "fear of all non-zero risks" and support for "scientific nerve and boldness." Is society willing to pay the necessary price for the "high adventure of science?" In reply he quoted the words attributed to Prometheus that are emblazoned in the Great Hall of the National Academy of Sciences: "I made them, the people, to have sense and to be endowed with reason."

The Academy Forum came at the high point — in part created by it — of public attention to the recombinant-DNA controversy. Almost certainly, it reinforced a trend from the quasi-regulation of the NIH guidelines to legal regulation under legislative authorization. Indeed, the fear of federal law that had dominated Asilomar was by now transformed into hope among molecular geneticists that "sensible" congressional regulation might stave off "irresponsibly repressive" legislation initiated by "hysterical" citizens in local communities. Norton Zinder of Rockefeller University, in response to a report to the Academy Forum by NIH Director Frederickson said, "I would like to support, and I am surprised that I am going to do so, the idea of having legislation, federal legislation with regard to recombinant DNA research. The proliferation of local option with different guidelines in different states and different cities can only lead to a situation of chaos, confusion, and ultimately, to hypocrisy amongst the scientists involved. I strongly plead that the government move ahead on this as rapidly as possible." The basis for Zinder's concern was personified by Mayor Vellucci, who attended a press conference during the Academy Forum. Even after his very considerable exposure in Cambridge, he made his position clear, "I don't know your language," he said, "I'm an American layman. I think you scientists have to buckle right down and begin to talk the way we talk in America." Communication was not necessarily achieved by exchanging words. Mayor Vellucci went to Washington, but his message had preceded him. Even as the Academy Forum was going through its paces congressional staffs were drafting bills for legislative regulation.

Toward Congressional Action

The public controversy had had its impact on the executive branch as well as the legislative. In November 1976, the Secretary of Health, Education and Welfare, "with the approval of the President," had quietly convened a special Federal Interagency Committee on Recombinant DNA Research. The committee included representatives of federal agencies, both those that support or might support recombinant-DNA research and those with existing or potential regulatory authority in the area. Chaired by the Director of the National Institutes of Health, the committee was charged "to address extension of the NIH Guidelines beyond the NIH to the public and private sectors." This charge was a logical progression of the process initiated by the Asilomar conference.

The first task of the committee was to convince all agencies likely to support DNA research to adopt the guidelines. The second was to review the powers of existing regulatory agencies as they might relate to recombinant-DNA materials and applications. By early March 1977, NIH Director Fredrickson was able to report to the Academy Forum that the roster of relevant federal agencies adopting the NIH guidelines was almost complete. On March 15 the Interagency Committee (Appendix V) communicated to HEW Secretary Califano its conclusion that no existing federal agency had statutory authority to deal with all problems raised by recombinant DNA and that new federal legislation therefore would be required. The objective should be to "establish uniform standards for such activities throughout the Nation."

To achieve this objective the committee recommended that primary regulatory responsibility should reside in the HEW secretary. The secretary would license both public and private facilities for research, oversee production or use of DNA-recombinant molecules, and register all individual projects prior to initiation. The secretary would also have authority to inspect facilities, to make environmental measurements, and take any "other steps to ensure safety." The secretary's enforcement authority would include the power to enjoin noncompliant activities if necessary. Disclosure of information, however, would be limited to give "appropriate protection" to proprietary and patent rights in industrial settings. In view of the need for "a single set of national standards," it was recommended that local regulatory action at state and substate levels be preempted.

On April 6, in remarks to the Senate Subcommittee on Health and Scientific Research, chaired by Senator Kennedy, HEW Secretary

Califano said, "there is no reasonable alternative to regulation under law. Only continued research, proceeding under strict safeguards, will tell us whether these restrictions must continue in force or whether they can be relaxed at some time in the future."* The secretary commented on the essentials of needed legislation as outlined by the Interagency Committee and described a bill proposed by the administration to realize the essentials. The proposed bill had already been introduced by Senator Kennedy on April 1, along with several reservations and a series of questions that the senator felt needed answers before the legislation could be formulated. These reservations and questions, together with testimony already presented before two House committees (Science and Technology, Interstate and Foreign Commerce Subcommittee on Health), laid out the issues to be confronted in the legislative process. That strong differences of opinion existed within the Senate on these issues became clear when Senator Kennedy's subcommittee staff completely rewrote the administrative version of the legislation. The new version substituted a regulatory commission to exercise the broad powers that the administration would have assigned to the HEW Secretary.

The major underlying issue thus was revealed to be the degree and manner of regulation. On the question of how much regulation was necessary extreme opinion ranged from none to total prohibition. Neither extreme view appeared likely to determine the legislative product, but proponents of each naturally continued their efforts to push the legislation as far as possible in their preferred direction. Between the extremes were those who favored some degree of regulation, at least initially that of the NIH guidelines extended to cover all recombinant-DNA research, production, and uses. Within this moderate group, the cleavage was between those who preferred to stay as close as possible to the NIH guidelines and those who favored both a new appraisal and a new regulatory focus.

The first view was approximated in a bill produced by Congressman Rogers and his House Subcommittee on Health and Environment. It placed responsibility in the HEW secretary, advised by a committee that the secretary would appoint and be required to consult on many matters. The Rogers bill also assigned important regulatory functions, subject to the secretary's approval, to local biohazards committees that would include public representatives. The alternative view was expressed in the Kennedy bill, which provided for a presidentially appointed commission, administered under the HEW secretary but not sharing its responsibility and authority with him. Such a public regulatory commission would clearly represent a departure from the self-regulatory concept that had been initiated by involved scientists prior to and at Asilomar.

* Statement of Joseph F. Califano, Jr., Secretary of Health, Education, and Welfare, before the Subcommittee on Health and Scientific Research, Committee on Human Resources, United States Senate, April 6, 1977.

Additional issues emerged in the legislative process. The original definition of recombinant DNA had been an operational one derived from
scientific usage. However, it was clear that whatever degree of hazard
might be inherent in recombinant DNA was not the result of the process
of recombination but of the properties of the two DNAs that were
brought together. Certain recombinants that duplicated those occurring
in nature would pose no *new* hazard or no hazard at all. Such recombinants needed no regulation and, if they were to fall under control,
would only increase the cost and the cumbersome nature of the regulatory process. Therefore, efforts were made to define recombinant DNA
legislatively to meet regulatory requirements.

Another issue involved the relation between federal and local regulatory authority. The Interagency Committee had called for complete
federal preemption of local authority in the interest of uniform national
policy. Senator Kennedy and others felt it important to maintain a local
capability for regulation, both because local circumstances might differ
and to encourage public participation in decision making that might
affect the public health and welfare.

A third issue centered on enforcement and the nature and degree
of surveillance and punitive measures required. Suggestions for punitive measures ranged from simple license withdrawal to threat of felony
charges and imprisonment. Yet another issue related to the protection
of confidential and proprietary information — whether the need of scientists for access to information or the property rights of commercial
interests should be protected.

Debate over all of these matters was complicated by controversy
and continuing uncertainty over the actual degree of biohazard involved in particular situations. It was also complicated by a half-hidden
agenda of social and ethical issues that were even more difficult to
resolve than the scientific and technical ones. In response to these
complications, both versions of the legislation provided for a comprehensive two-year study of all issues raised by recombinant-DNA research, including ethical and social considerations. In the Senate version this study would be carried out by the regulatory commission itself;
in the House version a special commission was provided that would go
out of existence on completion of its task. If such a study commission
were to be created it seemed that much regulatory detail, beyond that
already contained in the NIH guidelines, could be left for resolution
until completion of the commission study. In effect, the pending legislation would thus be an interim step, legitimizing and generalizing existing regulation while more definitive requirements were being assessed.

However valid the criticism may have been that the earlier phase of
policy formulation restricted public participation, the legislative history
included repeated hearings before four congressional committees: the
Subcommittee on Health and Scientific Research, which initiated legis-

lation in the Senate; the Subcommittee on Health and Environment, which initiated legislation in the House; the Science and Technology Committee, which held extensive oversight hearings in the House; and the Senate Subcommittee on Science, Technology and Space, which held its own oversight hearings in November 1977. Moreover, during the period of congressional consideration the subject was covered extensively in the public press. Attention never resulted in the intensity and generality of discussion generated by such matters as the breeder reactor, the neutron bomb, school desegregation, or abortion. Nonetheless, by the summer of 1977 ample opportunity had been afforded for public opinion to form and be expressed. What emerged were polar views, pro and con, and a fragmented middle ground of uncertainty and poorly articulated unease.

In midsummer, 1977, a new trend became apparent. During the preceding six months information had been accumulating that was leading to downward revision of risk estimates for most experiments with *E. coli* K–12 (see Chapter Six). At the same time alarm was growing in the biomedical community over what seemed to be an excessively punitive approach, localized especially in the Senate subcommittee. Scientific groups moved decisively to stem a tide that now conceivably threatened not only recombinant-DNA research but also basic research in general. The legitimate concern over biohazard was clearly energizing a deeper debate over the role of the public and government in influencing the directions of scientific investigation. A wider circle of the scientific community sensed danger and became involved in the legislative maneuvers.

Earlier in the summer of 1977 a large majority of attendees at three different Gordon Conferences had signed letters of concern. On July 28, Harlyn O. Halvorson, President of the American Society of Microbiologists, addressed a communication to all members of Congress on behalf of a coalition of professional biomedical societies. He noted that the coalition had been seeking to be responsive and helpful to public and congressional concerns but that it was "greatly disappointed" with the bill that had emerged from the Senate Committee on Human Resources, the parent committee of Senator Kennedy's subcommittee. A particular concern of the coalition was the committee's failure to take into account "new information showing that risk involved in recombinant DNA research has been vastly overstated and that recombinant DNA molecules do not increase the potential hazard of any microorganism." The dangers involved were regarded as "no greater than those encountered when dealing with natural pathogens." This statement was said to be supported by the fact that "no harmful effects or spread of recombinant organisms have been reported since recombinant DNA technology was introduced four years ago."

A delegation sponsored by the coalition, as well as a group of scientists brought together by the American Association for the Advancement of Science, underlined these and other important considerations in visits to key legislators and their staffs in both houses of

Congress. Other communications from constituents to key legislators
emphasized the growing concern of scientists and the importance of
recognizing that expert reassessment showed the risk to be considerably less than was thought to be the case at first. The progress of
congressional legislation was slowed, and no bill reached the floor of
either house of Congress. By early 1978 a new interim bill had been
drafted in the House, but Senator Kennedy had withdrawn his own bill
and joined with a group of other senators in calling on the HEW secretary to operate under existing statutory authority and extend regulation
to all recombinant-DNA research. Freed of immediate anxiety, legislators could replace their haste with deliberate caution. The panic of the
public's first look at the new-found powers of genetic intervention
seemed to be subsiding. The body politic was ready to react more
calmly and rationally to the double image cast by the double helix.

CHAPTER SIX

The Issue
of Research
Regulation

The course that led to regulation of recombinant-DNA research moved from recognition of need by individual scientists, to voluntary self-regulation within the involved scientific community, to quasi-regulation mediated by the NIH as a federal funding agency, and then closer and closer to full regulation enforced by a statutory authority with punitive powers. The visible driving force for this transition was concern about biohazard. Yet, curiously, the regulatory lightning struck the *uncertain* hazard of recombinant forms of DNA rather than, for example, the *known* hazard of pathogenic viruses. Did the public controversy over recombinant-DNA research stem entirely from the question of possible biohazard? Or did it originate in deeper concerns about the longer term implications and consequences of molecular genetic research? Did it go still deeper, expressing conflict over the social impact of all research? Will the recombinant-DNA issue prove to be a harbinger of broader public questioning of research in general? Answers to these questions are not clear, involving as they do the complex problems of interpreting the "public mind" and forecasting its future course. In this chapter we will explore the rationale for regulating recombinant DNA in relation to the larger issues raised by these questions.

Certainly recombinant DNA has special aspects. DNA is an unusual material and a crucial one. Within living organisms it is enormously rich in information, constituting, in the totality of contemporary organisms, a record of life's cumulative, successful adventures. Each generation, our own included, expresses the coded message of a particular descendant DNA that contains the distilled biological experience of an entire ancestry. The fact that these crucial coded messages have become accessible to biochemical manipulation has given scientists new ways to intervene in living things, not only in their present forms but possibly in their future evolution as well. This capability for intervention could conceivably be extended to the human species. Is public unease focused not so much on biohazard in the narrow health sense as on the less tangible consequences of genetic intervention, with the psychological and social tensions it generates? In this respect, the implications of recombinant DNA are not only unprecedented but they have reached a new scale of potential impact. The capability for seemingly simple human intervention in processes formerly thought to be subject either to blind exigencies of nature or divine will can easily stir anxiety.

Consider the situation in another way. There are many replicative DNA lineages — primates, birds, reptiles, insects, the major plant groups — and each of these forms of life contains a record in biochemical language of its success in its particular sector of biological experience. One lineage ascended through primitive vertebrates and primates to ourselves. This lineage evolved a second powerful mechanism for storing, processing, and reexpressing life's experiences. The mechanism began with the central nervous system. The functions of this system were first integrative, then cognitive, and now they underlie the production of human knowledge. Like DNA in biological heredity, knowledge is the selected and ordered record of past experience. It is transmitted from one generation to another, it is cumulative, and it is coded in social language. Human knowledge is a system for cultural heredity, as crucial in human social progression as DNA is in biological progression.

Intervention: Threat or Promise?

Molecular genetics has provided a possible means by which knowledge, as a peculiarly human form of heredity, may be used to manipulate and intervene in biological heredity. Moreover, since knowledge is accumulated and may be used to achieve purpose, the new intervention can be *purposeful,* displacing in part the slower natural mechanism of reproductive change. Is conscious and purposeful human intervention in the progression of life a threat or an opportunity? Is this new dimension for intervention, with its potential for doing either good or evil, the truly troubling ambiguity of the double image of the double helix? Is it possible that malevolent intervention in heredity, a Dr. Jekyll face of the double image, is causing distrust of the Mr. Hyde face, the potential biomedical benefits? Does this ambiguity, coupled with other anxieties about the impacts of knowledge-derived technology, provide the real thrust toward regulation of recombinant-DNA research?

The heroics of the Copernican revolution, the still vibrating shock of Darwinian evolution, and the early emotional resistance to vaccination are a few examples of social sensitivity and anxiety in the face of radically new knowledge. A world still not comfortable with atomic fission and adventures in space is now confronted with intervention in and manipulation of the innermost mechanisms of life. The notion is fascinating but also fearsome. Knowledge has always facilitated and stimulated intervention in nature's ways. But the scale of intervention now begins to rival in both magnitude and intricacy nature's forces themselves. Selecting seed to produce more abundant crops was an intervention. So, too, was harnessing the energy of fossil fuels. Experience shows that some impacts of these interventions extended far beyond original intentions or expectations. And now we stand at the threshold of new interventions with far greater implications for the human race.

Our knowledge and use of petrochemicals, pesticides, antibiotics, and nuclear fission brought benefits. They also brought a melange of unanticipated changes, some of which have not fully registered even yet. The public mind has experienced massive pollution of air and water, wholesale destruction of life forms, even the possibility that human intervention has changed the worldwide climate. Against this backdrop recombinant DNA has been visualized as a source of world pandemics and a new device by which tyrants might tinker with the innermost intimacies of life.

We are all learning that the decisions we weigh today will have consequences tomorrow that are beyond our ability to forecast. Genies are popping out of innumerable bottles. Our growing powers to intervene have made the possibility of self-generated harm almost a routine factor in the calculations of decision makers, whether the issue be weapons, energy, or contraceptives. Repeatedly we are forced to ask whether the wisdom of our purposes matches our power to intervene. We have wider and more attractive options, but our decisions involve greater uncertainty and anxiety. Our ability to predict consequences falls far short of our powers of intervention and our collective sense of purpose is puny in comparison to both. No wonder the fateful apple, the Faustian bargain, and the mushroom cloud are all part of the background of Asilomar, Cambridge, and the drive to contain recombinant DNA.

In these terms, the fear of biohazard has been only the cutting edge of the social impact of the recombinant-DNA issue. It is one side of a two-edged sword. Cutting one way, manipulation of DNA means deeper understanding of genetic processes and the bacterifacture of essential pharmaceuticals. Cutting the other way, similar manipulation might consciously or inadvertently increase the threat of disease or environmental danger. Whether and how we should continue this two-edged exploration provides the basic uncertainty. The stakes are high and the uncertainty generates anxiety.

Uncertainty is no stranger to human affairs, and it does not always induce anxiety. On the contrary, we exploit uncertainty for fun in our games of chance and we exploit it for military advantage in weapons technology. We also concede huge rewards to invested risk capital and give high emotional returns to those who bravely do battle "against heavy odds." It is the business of the gambler, the venture capitalist, and the soldier to "understand" the risks, to carefully calculate the probabilities for success and failure. Those who do well at risk-taking are usually not the repeated "plungers" but the methodical analysts who study each development and make their moves with both hindsight and foresight. To avoid all risk is to forego all opportunity. To be unaware of risk is to court disaster.

A New Assessment of Biohazard

A new assessment of the potential for biohazard in recombinant-DNA research has been developed since Asilomar. It seems important to restate the matter as it stands today, especially as it relates to the kind of regulation that now may be required. First, some degree of risk in particular recombinants is clearly conceivable, though neither how many such recombinants nor the degree of their possible harm can be specified. Whatever the degree of risk, it neither relates to the act of recombination itself nor is it equal for all products. Second, we must keep in mind that many genetic recombinations occur normally in nature. Recombination is unlikely to be dangerous just because it is performed artificially. Whatever the risk of recombining DNA may be, it lies in *particular* recombined DNA segments, not in the act of recombination or in all of its products. It also lies in the relationship of recombinant DNA to host, which conceivably can yield new and unexpected properties, different from those expected of either segment inserted alone.

Third, it is important to emphasize again that recombinant DNA poses little risk as an isolated material. No matter what the source, isolated and purified DNA in hand or on the shelf is inert both chemically and biologically. To display its significant properties it must be a functioning part of a living organism. Isolated DNA is not easily restored to a functional role. There are important limiting conditions. To say the same thing in another way, restoration of isolated DNA to biological expression is a low-probability event; it will occur only under special circumstances. This accords with the fact that life itself is a low probability event. Living organisms do not arise *de novo* in today's world, and they exist extraterrestrially, if at all, only in relatively special niches in the total universe.

Nonetheless, isolated DNA *can* be restored to function in a living organism. Moreover, life under natural conditions *is* tenacious and even today can be controlled only in limited ways. Therefore, it is not self-evident that particular artificial DNA-recombinant organisms escaping laboratory control may not propagate in natural populations. The quantitative probability can only be specified by actual experience or by technical forecasting based on related experiences. Whether any such successful escapees would have harmful effects on human or other populations similarly requires careful technical evaluation.

Fourth, growing experience of the past five years still allows experts to come up with different estimates of biohazard. However, over time the expectation of catastrophic hazard has declined among informed persons. The fact that concern is declining does *not* mean that actual hazard has changed. What has happened reflects the nature of scientific forecasting. More and more, successful science requires extreme specialization of effort. The process produces highly focused experts who are sometimes facetiously said to "know more and more about less and less." Molecular geneticists are particularly specialized because their field has been rapidly moving and is highly competitive. Members of this group first sensed risk in certain possible experiments and, as responsible human beings, called it to community attention. They are to be applauded for their sense of social responsibility.

However, assessing the probability of the escape and natural survival of artificial recombinant organisms is not the special expertise of molecular geneticists. Initially, no one was an expert in this area because the particular situation was novel. But related situations do exist — in the fields of microbiology, infectious diseases, epidemiology, ecology, population genetics, and evolution. Gradually, the attention of experts in these fields has been attracted to the recombinant-DNA issue, and these experts have now contributed to sounder assessments. Their appraisal is generally more conservative with respect to the possibilities of extreme biohazard than the earlier one made by the molecular geneticists. The reasoning of these other experts does not eliminate all possibility of risk, but it tends to restrict the possible risk and make it appear more manageable.

The new consensus stems from the following considerations. The record of research with dangerous microorganisms that cause such diseases as cholera or plague shows the need for great caution but does not support the likelihood of catastrophic spread of disease *from laboratory foci.* Experimenters and their assistants have become infected with significant frequency and historically illness and death have not been uncommon in this group. With growing experience and technological improvement, however, these unfortunate occurrences are declining. In addition, what is judged particularly significant is the rarity of occurrence of secondary spread, that is, infection of persons not directly associated with the laboratory. Epidemic spread from laboratory foci has never been recorded.

This information, unfortunately, is not applicable in a simple way to the forecasting of potential biohazard of recombinant DNA. Known pathogens, on the average, are likely to have more immediately visible effects than recombinant organisms. For that reason they are recognized relatively early and appropriate known control measures can be applied. Recombinant organisms might require greater time to give evidence of pathogenicity and might, by then, be widespread.

On the other hand, most known pathogens require a long biological history to achieve stability in their parasitic way of life. Infectious disease experts say that the capability for successful pathogenicity is not easily achieved. Successful pathogens must have ways of getting into potential hosts and of overcoming their strong resistance against invasion. Once inside, pathogens must have mechanisms not only for growth and reproduction within the host but for survival while transferring to a new host. Each step would be expected to require a new complex of genes. Clearly, such necessary combinations of genes have been evolved, over time and under natural conditions, by existing pathogens. But to create these new sets of properties by artificially adding individual DNA sequences to an already large hereditary complement would not be easy. It is relevant that despite intensive efforts to develop new pathogens in the biological warfare experiments attempted during and after World War II, no striking new pathogens appear to have emerged. This argues that the artificial production of transmissible pathogens is not likely to occur inadvertently or easily. More likely, it is as difficult as the artificial production of any other complex biological property.

Such considerations lead many experts to conclude that the likelihood of *accidentally* producing dangerous new pathogenic parasites in properly controlled recombinant-DNA experimentation is not high. By reference to these same considerations, the possible *deliberate* production of pathogens by concerted recombinant techniques in warfare, terrorist activities, or other irrational pursuits is conceivable and not easily evaluated at present. It is small comfort to point out that many other technologies are similarly vulnerable, for example, those involving chemicals, radioactive materials, and explosives. Such risks of technological misapplication can be minimized but not eliminated. In a society that maintains maximal openness in all of its activities a reasonable degree of surveillance would appear to be the best answer.

Some argue that the risks are greater for recombinant organisms than for other technologic threats because recombinants are "new forms of life" that, once released, can propagate without control. Recombinants are, however, much less new forms of life than they are modifications of existing ones. Moreover, microbiologists and ecologists note that scenarios of "wild" and uncontrolled propagation of dangerous recombinants do not take into account information of two kinds. Growing evidence shows that the transfer in nature of DNA segments among "unrelated" microorganisms is much wider than previously suspected. Novel combinations may, therefore, be more frequent than was earlier believed. This also means, however, that artificial DNA recombinants, once introduced into natural populations of microorganisms, have a greater likelihood of spreading widely. Fortunately, recent studies show that artificially introduced DNA segments usually reduce rather than increase the competitive success of the altered host. This is not surprising to ecologists, who know that adaptive success of organisms in natural environments is subject to many checks and balances. For example, genes that add to energy requirements without conferring a specific adaptive advantage are speedily eliminated from a population. Overall, then, the more recent and more comprehensive scientific assessments of the risk of recombinant-DNA organisms present a somewhat less alarming picture than the earlier more limited ones.

A letter from Professor Roy Curtiss to NIH Director Donald Fredrickson on April 12, 1977, makes this point. Curtiss points out that in 1973 he was more concerned about biohazard associated with recombinant DNA than was the Berg group, which called for a moratorium on *certain* recombinant-DNA experiments. In an open letter written at the time, Curtiss directed attention to additional factors for concern and suggested a voluntary cessation of essentially *all* recombinant-DNA research pending further assessment. Subsequently, at the Asilomar conference and as a member of the NIH Recombinant DNA Molecule Program Advisory Committee he continued to urge maximal caution until potential biohazard had been more fully explored.

In August 1974, Curtiss' concerns led him to discontinue all his own efforts at constructing recombinant DNA in order to give his full attention to developing safer K – 12 strains. He also undertook a program of self-education in all areas of expert knowledge pertinent to assessing and improving the general assessment of the biohazard involved in recombinant-DNA research. In his 1977 letter to Frederickson, Curtiss reviews the status of his assessment at the date of writing and states his resulting views. He finds it "difficult to believe," on the evidence, that *E. coli* K – 12 could be converted into a pathogen by introducing foreign DNA sequences. Even if a gene for a poison were introduced the only individual at risk would be someone who swallowed large quantities of the culture, either carelessly or deliberately. Communicability of the infection is very unlikely. "Even if there were a natural catastrophe such as caused by an earthquake, tornado, hurricane, etc., it is unlikely that *E. coli* K – 12 containing recombinant DNA could initiate or sustain an epidemic. . . ." With respect to the communication of DNA among microorganisms in nature, now believed to occur quite commonly, he considers the transmission of recombinant DNA introduced into K – 12 on suitable vectors to be "a most unlikely event," and even less likely with the genetically crippled strains he and others have developed to meet EK2 requirements.

Curtiss draws the general conclusion that "the introduction of foreign DNA sequences into EK1 and EK2 host-vectors offers no danger whatsoever to any human being" other than to a particular individual who carelessly or deliberately exposes himself or herself to specially pathogenic strains. He notes that "the arrival at this conclusion has been somewhat painful and with reluctance since it is contrary to my past 'feelings' about the biohazards of recombinant DNA research." He carefully excepts from his conclusions the possibility that deliberate attempts to construct a pathogen might succeed "with considerable skill, knowledge (most of which is currently lacking) and luck." It should also be noted that Curtiss' painstaking efforts and conscientious appraisal apply only to the *E. coli* K– 12 host-vector systems.

Participants in a special workshop on risk assessment, conducted with the support of the NIH at Falmouth, Massachusetts, on June 20– 21, 1977, concurred in Curtiss' general views. The workshop included some fifty invited experts on various aspects of *E. coli* biology and pathogenicity, selected without respect to their involvement in recombinant-DNA research. Having reviewed past and recent information, including reports of deliberate efforts to convert *E. coli* K– 12 strains into pathogens, they reportedly were unanimous in their opinion that such strains cannot be made into dangerous pathogens by DNA inserts.

All basis for concern is not removed by these findings and opinions. However, under the precautions required by the NIH guidelines, the probability of accidental production of *E. coli* pathogens and their epidemic transmission is clearly now assessed *much* lower than before. While there is no absolute certainty that dangerous *E. coli* recombinants may not be produced either intentionally or inadvertently, the likelihood of them becoming established in nature by chance is judged to be vanishingly small. Even with positive effort and planning there is reason to doubt that it could be achieved. Therefore, potential biohazard from *E. coli* recombinants requires continuing attention, but the level of immediacy is not on the order of a flash fire in a drought-ridden forest. What seems desirable is suitable regulation as part of the cost of venturing into a still largely unexplored area. Curtiss and others caution against "excessive regulation ... based almost entirely on fear, ignorance and misinformation."

Genetic Intervention and
the Regulation of Knowledge

If estimates of potential biohazard do not justify extreme regulatory steps, is concern about genetic intervention sufficient to tip the balance? Manipulation of heredity has long been a practical goal and yet has inspired concern. Emotional controversy has erupted in the past when

new means for such manipulation were developed. Many people cannot
forget the debate over eugenics in this country early in the century and
the tragic consequences of misapplied hereditary theorizing in both
Nazi Germany and the Soviet Union. Concerns now have been rekindled by the apparent relative ease with which we might restring the
beads of hereditary fortune or misfortune.

This kind of intervention in purely *physical* or *chemical*
processes—for example, energy transformation or invention of new
plastics—is viewed by our culture with greater equanimity. Physicists
and chemists are applauded for feats of material invention and transformation far beyond the dreams of the alchemists of centuries past. But
biological alchemy is another story; intervention in life strikes deeper
fears. The highest resistance, of course, comes with suggested intervention in the intricacies of human beings, especially as these approach
personal individuality. The notion of human "genetic engineering" is
especially fearsome. It contains the threat of insidious and powerful
penetration to the vulnerable inner self. It bends the security prop
provided by the concept that humanity is stable even if individuals are
mortal and transient. Despite the fact that such interventions in their
most dramatic form are far down the road at best (or at worst), the fear
inspired by the possibility is psychologically, philosophically, and socially significant today. Almost certainly this fear feeds the deeper anxiety that goes beyond the fear of biohazard and motivates the notion of
regulating recombinant-DNA research.

This special aspect of projected risk leads to penetrating questioning of purposes, motives, and values. Is the opening of this door justified by the consequences, known and unknown? The involved scientist answers readily: Open the door very carefully and see what lies
beyond. Responses by uninvolved people are more cautious. To stop
and consider all risks, all benefits, and all available alternative courses
seems a reasonable approach. These reactions and intermediate combinations of them constitute the substance of the recombinant-DNA
debate. In the debate the movement from self-regulation by the scientific community towards external regulation under law has raised a
number of deep issues. We must fully understand where the regulatory
road may be leading. It generates a new concern. Something more
fundamental than even fear of genetic manipulation may be
operating—the fear of knowledge itself.

The Risk of Controlling Knowledge

Regulation is now very much part of everyday life. We regulate air travel,
the stock market, automobile exhaust, food and drug quality, even the
number of and equipment in restrooms in public buildings. Much of
this regulation is justified and necessary but some has proved cumber-

some, expensive, and counterproductive. In recent years, both regulation and the resentment of regulation have been highly visible and the focus of political controversy.

To date, the pursuit of fundamental knowledge in the United States has remained largely free of direct official regulation. Academic and scientific freedom are bound into the American democratic heritage. These liberties are defended by impassioned rhetoric and maintained by a reasonably stable tradition. Fundamental science has been included in the sacrosanct zone with only an occasional challenge. Now, however, the situation may be changing. As the relationship between basic science and technology has grown stronger and technology has come under fire, basic science has received suspicious scrutiny and some of its blemishes have attracted attention. Regulation of recombinant-DNA research would not be the first instance of formal legislative regulation of scientific research in the United States. An earlier example was regulation of the use of human subjects in experimentation.

The connection between these two sorts of activity is not accidental. Experiments with human subjects have been conducted for many useful purposes for a long time. In medical research, no matter how many experiments are done first on animals the final test of a drug or a procedure must be carried out on humans. Moreover, many medical and behavioral questions are peculiar to humans. For example, only background information can be obtained from animals about many aspects of human intelligence or human emotions.

Some experiments on human beings were done under less than satisfactory conditions, and in 1974 the use of human subjects came under formal and stringent regulation. The National Research Act (P.L. 93-348) requires that all "biomedical and behavioral research involving human subjects" be reviewed by an Institutional Review Board. The purpose of the review includes the protection of the rights of the subject. Whether fundamental or practical, experiments are proscribed if they fail to protect such rights. The regulations that implement the legislation make clear that the review should determine whether "the risks to the subject are so outweighed by the benefits to the subject and the importance of the knowledge to be gained as to warrant a decision to allow the subject to accept those risks."

The principle thus established is that threat to human rights takes precedence in experimentation unless overbalanced by *substantial* benefit. Preferably, the benefit should devolve directly to the subject, but it also may manifest itself as new knowledge advantageous to the whole human race. The sum of both sorts of benefits must be greater than the risk. Moreover, the subject must understand the risks and benefits and give informed consent. Although the principle is cast in formalities and legalisms, the motivation clearly lies in the desire to avoid violation of the sanctity of the individual. Experimental intervention cannot override self-determination.

This principle is being applied and extended in the regulation of recombinant-DNA research. A particular *research area* rather than a human subject is brought under regulation, with the intention of protecting the health and other unspecified rights of the *public*. Again, the assumption is clear that protection of such rights takes precedence over research objectives unless these rights are overbalanced by substantial benefit. The benefit may be experienced by the public either directly or indirectly through the accrual of significant and useful knowledge. With respect to the regulation of recombinant-DNA experimentation, provision is made for public participation, a kind of public "informed consent." Is this extension of the principle of risk-benefit calculation and informed consent, adopted first to protect human subjects of experimentation, appropriate for research in general? Do all experiments put the public at risk? Has the transformation of the directions and organization of science put the public at risk more often and more significantly than in the past, thereby justifying legislative regulation? These questions go to the heart of the changing relationship between society and the generation of knowledge.

In simpler times, the scientific community regarded *any* restriction or regulation of research as unacceptable if it impeded or distorted the search for knowledge. However, World War II constituted a kind of turning point in the relationship of science and society. Since midcentury, fundamental science has become almost entirely dependent on public funds. At the same time public decision and action have become ever more dependent upon knowledge and the contributions of science. Moreover, the Nüremberg trials that followed World War II made clear that the principle of unrestricted inquiry, especially in the use of human subjects, could lead to shameful and repugnant abuse. Clearly, nothing—including the claim to new knowledge—justified the interference with essential human rights. Over the succeeding quarter-century, restriction backed by external authority became increasingly acceptable as a necessary precaution to eliminate from experimentation clear danger to human rights.

Complications arise, however, even in interpreting the original doctrine and certainly in extending it. The Commission on Use of Human Subjects in Biomedical and Behavioral Research has been examining the difficulties of interpreting informed consent with respect to, for example, fetuses, children, the mentally incompetent, prisoners, and other special groups. Applying protection in recombinant-DNA research now requires a definition of *public* informed consent and other public rights. The voluntary moratorium of 1974 assumed that possible public exposure to new dangerous pathogens justified a proscription of certain recombinant-DNA experiments. The NIH guidelines confirmed this assumption, and the legislation discussed in the last chapter sought to give the proscription the force of law. Suppose, however, that these same experiments could illuminate the nature of disease-producing organisms and afford more effective means to control them. By prohibiting this research would we be denying rights to those whose lives might be spared? Who should weigh these questions and how should such decisions be reached? Are the decisions scientific, ethical, political, or some combination of all three? How adequate are our social mechanisms for dealing with questions of this kind?

These problems are difficult but not entirely new. They were raised, for example, as the "Mondale questions" in the legislation that created the Commission on Use of Human Subjects and they remain on the agenda. Similar questions come up in relation to a wide range of other public-policy decisions. With respect to recombinant-DNA research, consideration of these questions highlights three significant factors that need careful consideration: (1) the nature, degree, and means of assessing risk; (2) the general value to be assigned to new knowledge; and (3) the specific measurable values generated by the practical application of knowledge. Sound public policy for expansion of knowledge demands appropriately balanced attention to these factors.

Risk and the Need for Regulation

Dealing first with risk, it is clear that certain research risks may be as direct and obvious as the risk of fire or highway accident. Administration of a suspected cancer-inducing agent to a human subject or release of a new known pathogen into a community environment are examples. Such experiments clearly require previous analysis and disclosure. Either safety must be demonstrated or the necessity for the risk and the precautions to minimize it must be accepted. Individuals or populations placed at risk must be informed and must freely give consent through an established political process. Where the risk is statistical, in that no particular individual at risk can be identified, the political process must ensure public awareness and appropriate public participation in the decision-making process. We do not yet have effective mechanisms for eliciting from either a select or general public a reasonable equivalent to individual informed consent. This is the heart of the public-participation issue raised in the recombinant-DNA controversy. It urgently needs careful study. Public participation was reasonably effective in the Cambridge episode. It is not yet clear whether it will operate with equal effectiveness at the national level.

Risks other than biohazard are even more difficult to pin down. Some require an evaluation of trade-offs, and they may be so long-term that effects will sometimes only register in descendant generations. The risk of excessive alcohol consumption by pregnant women or the relative benefit of saccharin to the general population and to diabetics fall in this category. So, too, do experiments on the effect of different forms of mothering on newborn infants or the possible effects of recombinants that include DNA from potential tumor viruses. In none of these situations is harm to a given individual a certainty, but the possibility of statistically undesirable effects within a population can be foreseen. The two experiments cited, however, may also yield benefit over the longer term to many individuals with real problems or illnesses.

The complexities of this second category of risk require expert appraisal and may not yield unanimity even then. Analysis and decision making are especially difficult because in these matters value judgments are as important as scientific considerations. In such circumstances there is real danger that ill-considered regulation paradoxically may cut off the very research and additional knowledge necessary for sounder assessment. Regulation, therefore, must be applied with particular caution. In general the burden of proof in this category would seem better placed upon the regulator, with the benefit of doubt given to the experimenter. But every case must be examined for itself, at least until a body of effective precedents has been established.

The third kind of risk is entirely based on conjecture or speculation, and is often seen as both indirect and long-range in its effects. At the time the risk is assessed, there may be little or no objective information. A case in point is the controversy, discussed earlier, over recombination of DNA from bacteria and higher organisms. The suggestion that such experiments constitute a dangerous violation of evolutionary barriers is alarming, especially to a public mind already sensitized and made apprehensive by the adverse effects of earlier technological decisions. Such speculative scenarios for disaster can have their usefulness and must be carefully considered, but they should not become the basis for regulation without concrete positive evidence. The fact that such evidence is not easily attained does not mean that speculation should be turned into an operative assumption. This sort of speculation in particular feeds public distrust and threatens the expansion of knowledge, especially when it is coupled with the suggestion or assumption that scientists and technologists are often driven by self-interest and irresponsible zeal to perform activities that are dangerous to the public.

It is important to distinguish among these categories of risk situations because each requires a different regulatory reaction. It is true that in experimental research, as with exploration in general, consequences (risk included) cannot be fully predicted. If this were not so there would be no point to doing experiments. Despite this limitation, however, the first risk category—possible biohazard involving direct harm to people—is most amenable both to risk assessment and reduction. The third—risk as determined by speculation—is least amenable to regulation, and the second—where subsequent rather than current populations are at risk—is intermediate. To discuss all three simultaneously can only lead to loose, confusing generalizations about risks and benefits. Clearly, for risk-benefit calculation to be determinative, one must be able to quantify both the risks and benefits and in commensurable terms. Obviously, the risks in the three categories described cannot be quantified to the same degree. In the third case discussing risk in these terms is essentially impossible. Thus risk-benefit analysis is meaningless for such totally speculative scenarios.

Possible Benefits and the Need for Regulation

In turning to benefits, the other element in risk-benefit analysis, a distinction must be made between research intended for use and research meant only to increase knowledge. So sharp a distinction is not desirable in some other contexts. For example, when we trace the flow of knowledge from basic biomedical studies to the bedside it is more important to emphasize the continuum than any distinction between the two ends. In a regulatory context, however, making this distinction is essential.

The distinction between "pure" and "useful" research is necessary because quantifiable benefit only appears as knowledge approaches use. Increments to knowledge are inherently beneficial if only because they satisfy a basic human impulse toward new experiences and expanded overall perspective. In turn, these broadened perspectives become practical factors in newly perceived choices, and background factors for decisions. But it is usually impossible to assign to such new knowledge a precise quantifiable value in terms commensurate with risk. For example, what is the value of the concept of a black hole or the first footstep on the moon? It is only when we see a clearly defined practical product or result that a number can be applied — whether in dollars, lives saved, or ergs of energy released. Knowledge not related to use is a clear benefit but it is a *nonquantifiable* one. Nonetheless, such qualitative benefits must be reckoned in calculating the effects of regulation. Indeed, special status must be assigned to them, both because of the nature and importance of new knowledge and because of the special character of the setting within which such research can go on.

The special nature of basic research, as discussed earlier, does not confer upon it immunity from all regulation. For example, clear and present danger to essential human rights cannot go unchecked. Biohazard of recombinant DNA is potentially in this category, which is why the controversy about its actual risk is so focal. The controversy can only be resolved by better information about actual risk and its mitigation.

However, when direct and obvious threat to clearly defined human rights is *not* involved the situation changes. Now objective risk-benefit assessment becomes more difficult and subjective value considerations play an increasing role. We may learn more about cancer and its management by probing for and isolating unusual DNA segments in tumor cells. We may also in the process release or concentrate cancer-inducing factors. Is experimentation in these directions more likely to expand benefits or to threaten public rights? Who carries the burden of proof?

Is the investigator personally liable if he or she makes a misjudgment?
Are these matters properly subject to regulation or should we continue
to rely on trial and error? This is the difficult area in which we are still
seeking sound and efficient decision-making mechanisms.

In the third category of risk, based on speculation for which there
is no evidence, the burden of proof clearly appears to shift to the
challenger rather than the researcher. The nonquantifiable but time-
tested benefit of new knowledge must take priority over risks that are
hypothetical and equally nonquantifiable. This conclusion applies even
more forcefully to concerns about *possible* intervention in heredity in
ways that *may* lead to evolutionary control. As noted earlier, molecular
genetics is part of a rising curve of intervention that began centuries ago
and has been steepening. Recombinant-DNA techniques are likely to
reinforce the trend. There is, however, no *a priori* reason to fear one
means of intervention more than another. Nor does available knowledge necessarily result in intervention. We selectively breed dogs and
horses but have successfully resisted suggestions to deliberately breed
for human characteristics. It is not means but consequences of intervention that appear to be the subject of legitimate concern.

Can Science Be Trusted?

Here a conception dangerous to the effective cultivation and the proper
role of knowledge is being propounded by some. It is argued that a new
direction of research can never be fully assessed as to consequences
and always involves some risk. The risk inevitably is borne not only by
the experimenter but by society in general. Moreover, it is increasingly
difficult to assess consequences in a complex interactive society and, it
is argued, the motivations and purposes of the experimenter are not
always trustworthy. They can never be purely scientific since scientists
are people and have political and social biases. Research, therefore, is
always in part a political act and must be examined and regulated at its
very inception — at the point where its subjects and directions are chosen. Since the public is at risk, it has an interest and must play a role.

This argument has enough substance to be accorded serious status.
Many scientists are so deeply suspicious of the entire issue that they
bitterly oppose any discussion of how research directions actually are
chosen. This will not make the argument go away or reduce whatever
cogency it may have. Recombinant-DNA research is as good an arena as
any to cope with the issues raised. The fact is that the suggestion that
scientists are socially conditioned in their behavior is intuitively reasonable, is testified to by many scientists, and fits the observations and
interpretations of historians and sociologists of science.

It is equally true, however, that scientists are conditioned in
another way as well — by interaction with the real world of phenomena.

The philosophical issue of the nature of reality neither interests nor disturbs scientists as they go about their daily business. They know that a large factor in their social conditioning and bias are the attitudes commonly held in the scientific community. Especially important are attitudes and objectives of their coworkers in their particular branch or field of investigation. As a group scientists are focused on commonly recognized phenomena that have arisen in their collective experience. Whatever the other biases of individual scientists, their rewards are dependent on the satisfaction of colleagues as to the importance of the experiments they choose and the cogency of their results and interpretations. The "real world of phenomena" to the scientist at the bench is the commonly conceived world of a field of investigation. Appropriate new questions or directions of research are determined in that "inner world" of science, with some but often minimal interaction with the "outer world" of social needs and attitude. The question raised is what the relative importance of these two orientations should be and how they should function relative to each other in choosing research directions.

As research has penetrated ever more deeply into the phenomenal world each field of investigation has come to be like one working party in a vast complex coal mine. The working party knows its own mine face like the back of its hand and cannot understand how anyone else can judge better how to make that face produce. If the mine is to be productive the richest faces must be worked to their maximum. Only those working on actual faces can detect obstacles, sense dangers and determine how to overcome them. In these terms, there are no faces too dangerous to work, no phenomena too dangerous to understand. The task may involve risk but if society wants coal it must rely on the decisions of skilled miners to get it.

The perspective is by no means fully accepted by the mine managers or by those who finance the operation. Viewing it as a whole they may perceive more danger or greater productivity in one sector of the mine than in another. This has happened with recombinant-DNA research, leading to the reluctant acceptance of some degree of regulation by the miners. This acceptance is limited to concern about immediate danger but does not extend to management decision as to what faces are likely to be most productive. Most especially it rejects management's decision that certain faces may not be explored at all because of dangers that are vaguely specified and are not inevitable consequences.

Can Understanding Be Dangerous?

Departing from the analogy, it is reasonable to argue that there are *no* phenomena dangerous to *understand,* even though the path to understanding or the activities that follow upon understanding may involve risk. With respect to such subsequent activities an important distinction

must be maintained. The immediate and only *necessary* consequence of
knowing is understanding. And understanding is always a benefit which,
without subsequent decision and action, can pose no risk. The contrary
may be the case, however; not understanding may pose risk when
circumstances must be faced that require the particular knowledge.
Moreover, once understanding is achieved, actions based upon it require additional decision. These *uses* of knowledge are subject to
choice both by individuals and society as a whole. Social choice is a
political process and is subject to regulation in the public interest. The
only limitation on this sort of regulation is that it must not prohibit
knowing itself. Such prohibition risks forfeiture of important benefits:
the satisfaction of acquiring new knowledge, the advantageous use of
knowledge, and the further expansion of the human outlook. These
losses are too high a price to pay for the sake of allaying anxiety over the
unknown; it means cutting off the human tradition of sapience, that has
steered our biological and cultural course since the species came into
being.

Therefore, regulating the *uses* of knowledge is a legitimate re-
course, but regulating knowledge-generation itself is fraught with peril.
The central principle is that the search for knowledge must always
respect the highest human rights but the uses of knowledge must respect generally accepted human purposes. This principle must have
been recognized very early in human history. Regulation of uses had to
appear with the exploitation of fire. Even though fire is tamed, the uses
we put it to need regulation even now. Our knowledge of fire and all
other ways of releasing and using energy have grown apace; so, too has
the complex of regulations governing their use. If our application of
knowledge outpaces necessary regulation we may be in trouble, but if
regulation outpaces and cuts off the growth of knowledge, we are
metaphorically — and possibly in reality — dead.

Requirements for Sound Regulation

What then are the major considerations when circumstances demand
some regulation of knowledge and its uses? The first is the recognition
that expansion of knowledge should be regarded as one among the
highest set of human values and rights. Commentators on the
recombinant-DNA controversy have discussed whether freedom of research is guaranteed by the First Amendment of the United States Constitution. Scholars of constitutional law note that though research is not
referred to specifically in the amendment, both the language and its
judicial interpretation in other situations seem to apply to freedom of
research. Moreover, it can legitimately be argued that freedom of research and inquiry is founded in a "higher law" than the Constitution. It
is part of a fundamental strategy developed by the human species and

manifested in cultural progression over some 3 million years. A com-
mittment to accumulation of knowledge and, later, to deliberate inquiry
to augment accumulation is at the foundation of the growth of culture.

There is a case to be made, therefore, that the right to expand
knowledge belongs at the apex of any contemporary ethical system,
along with, for example, the sanctity of human life and the dignity of the
individual person. It is, however, only one of a *set* of such apical values.
As one value in that set, efforts to increase knowledge are subject to
limitation only by interactions *within* the set. The right to inquiry and to
seek knowledge is not an absolute. It cannot, nonetheless, be lightly set
aside even in deference to another apical value. Thus, as noted above,
research may not be carried out on human subjects without the most
scrupulous attention to the dignity of the human person. And informed
consent with voluntary acceptance of fully explained risk is an appropriate limitation of research on humans. On the other hand, human
dignity may not be invoked in ways that make all human experimentation impossible. There is a value in obtaining greater understanding of
ourselves that cannot be totally sacrificed to any other value, even one
in the apical set.

Where a *public* rather than an *individual* apical right is involved
the collective equivalent of informed consent must be defined and
achieved. It must be clear that any restrictions that result should relate
to the manner of inquiry and not to the objectives. Informed consent is
required for a surgical operation on the human brain because the
procedure carries risk, not because we ought not to intervene in or
understand the workings of the human brain. The *procedures* of research activity are limited by ethical considerations, not the objective,
which is the expansion of knowledge.

A second major consideration in the regulation of knowledge is
that its uses, as distinguished from the procedures for acquiring it, are
subject to general scrutiny and to limitation if circumstances clearly
require it. Unlike the value of knowledge and the right to expand
understanding, utilization of knowledge does not necessarily belong
under the apical set of ethical principles. More often, utilization involves
lesser values and frequently conflicting preferences within the body
politic. The controversy over birth control technology is a case in point.
Such conflicts are the proper subject for political mechanisms of choice.
In the operation of political choice ethical considerations have their
place as do knowledge considerations, and especially scientific knowledge. The more difficult the choice the more important it is that knowledge be brought to bear in evaluating options and consequences. In
controversial areas, therefore, sound regulatory policy will not limit the
generation of necessary new knowledge but encourage it, subject only
to such constraints as are necessary to protect human rights and generally accepted requirements for the public welfare. This is the strongest
agrument against those who, in their anxiety over possible health

hazards of recombinant DNA, would have clamped a general

moratorium on the whole subject.

Third, if measures to regulate research are necessary, they must be carefully drawn so they neither inadvertently nor deliberately extend beyond the minimal requirements. This danger is inherent in hastily made decisions or log-rolling compromises. In carelessly legislated regulation the distrust of knowledge itself may exert a significant influence, thereby inhibiting the generation of knowledge and cutting off an essential resource to deal with mounting human problems.

Fourth, significantly different kinds of knowledge-related activity require different regulatory approaches. The recombinant-DNA issue illustrates three general kinds of such activity: basic research, applied research and development, and manufacture and use. Basic, or non-use-oriented, recombinant-DNA research is largely carried out in academic or other nonprofit settings. It is usually federally supported and has been regulated since the advent of the NIH guidelines. Traditionally, such basic research is motivated by the desire to gain knowledge, it is self-regulated by peer pressure to assure standards, and it is maximally sensitive to such sanctions as withdrawal of external support or of licensure. Under these circumstances, control of biohazard or other risks can be achieved by rational discussion of potential dangers — combined with relatively simple and largely self-monitored surveillance by the experts already involved. Such an approach does require organization, particularly with respect to monitoring, as well as a reversal of typically casual, laissez-faire academic attitudes. Such activities are the assigned responsibility of the institutional biohazards committees called for in the NIH guidelines, and the activities of these committees must be carefully scrutinized to assure their effectiveness. Heavy-handed punitive measures borrowed from the bureaucratic handling of governmental-industrial tensions are not appropriate in this setting. Such measures are more likely to drive investigators out of the field than to achieve the balanced social objective of minimizing risk while allowing knowledge expansion to go forward.

The second kind of activity is applied research and development specifically oriented toward the practical use of recombinant-DNA techniques. This sector is likely to grow rapidly in the near future. Such activity is likely to involve some academic investigators, but more characteristically it will be conducted in industrial organizations. In this setting knowledge considerations are less dominant than financial objectives. Industrial research is an economic investment and its costs necessarily must be compensated by a greater net profit. Accordingly, early pay-off is generally emphasized, findings are owned privately and not placed in the public domain, and financial success outweighs scientific prestige. The motivation in industrial activity is the transformation of knowledge gained into products and economic uses. In the process ways are found to overcome obstacles, including governmental controls. Thus, industrial research and development provides a very different target for regulation than research in academic institutions.

Industrial research and development of recombinant-DNA technology is the central target for regulative consideration at this time. Little searching discussion of the policy issues raised has taken place. For example, are we ready for industrial application when considerable uncertainties still exist not only as to safety but also as to the potential and desirability of direct genetic intervention? Should research that is still of general scientific importance be kept under the cover of proprietary interest? We should also consider whether the large public investment that gave rise to recombinant-DNA techniques should go uncompensated while the techniques are put to use and acquire economic value. These are grounds to move slowly in commercially exploiting the new technology until deeper policy studies are carried out. In the interim, development of practical recombinant-DNA technology might best include continuing strong federal involvement.

Such considerations need not result in the cessation of industrial activity. Licensure and federal contracts and grants might be used to give impetus to industrial participation that would yield entrepreneurial advantage for later development and use. The focus of industrial research might be on issues of safety, particularly to reduce uncertainty as to risk and to develop alternative host-vector systems. It would be necessary, however, to require industrial researchers to report their work freely in the open literature. Though the research results would thus fall into the public domain, cumulative "know-how" would remain a competitive advantage, motivating the research initially and potentially resulting in longer-term financial return.

The third category of recombinant-DNA activity that must be distinguished in relation to regulation is production and use. This activity is probably farther down the road. Still, at this point it must be recognized, analyzed and considered in terms of policy planning. It is very likely too early to anticipate precisely either the nature of the recombinant organisms or the products that may be produced on a large scale. Production, however, will certainly pose different regulatory problems from research and development. The activities may be of many different kinds and may occur in many different locales, not only in the medical context but in agriculture, energy production, and even environmental modification and control. The potential breadth of foreseeable application thus exceeds the purview of a single existing federal agency. This diversity needs to be taken into account in designing appropriate regulatory control. A lead agency or an appropriate coordinating mechanism may be an essential element.

The recombinant-DNA debate so far has failed to distinguish among these differing regulatory requirements for basic research, applied research and development, and production for use. Legislative efforts have been dominated by a probable overassessment of immediate biohazard, by suspicion but too little hard information as to commercial intention, and by an inadequate definition of the possible

deeper issues. The immediate task is assessment and appropriate control of whatever biohazard exists and extension of such control to all recombinant-DNA research and use. Beyond that, a mechanism is needed for deliberate and careful study of future courses in the light of unfolding experience. The effort to foresee and solve all problems legislatively in quick and summary fashion not only may fail to solve the problems but may abort or distort many of the extraordinary potentials of our new genetic knowledge.

To reiterate, appropriate consideration of regulation rests on three assumptions: (1) Such risk as may exist must be soberly assessed in the most comprehensive terms and in a manner fully accessible to the public; (2) continued expansion of knowledge and the practical benefits this may generate is essential to the public interest and it should be restricted only to the degree that there is compelling evidence of clear and present danger; and (3) any regulatory action taken generates its own risk — of heavy administrative costs and bureaucratic inhibition of the search for knowledge. Recombinant-DNA research is a bellwether issue in this regard. Both impressive risks and impressive benefits can be reasonably projected, but the result cannot be predicted without clairvoyance beyond current forecasting capability. A number of interest groups have claimed such clairvoyance but the assumptions they make do not necessarily conform to those of total public interest.

Regulation and Interest Groups

Since scientists directly involved in the research necessarily feel threatened by the public debate and the possibility of external intervention, many of them tend to emphasize the lower range of risk estimates and the higher range of benefit estimates. Scientists not involved in recombinant-DNA research tend to support their colleagues because they are concerned about extension of external regulation to their own research areas. A minority of scientist-critics are on the opposite side of the debate for a variety of reasons, some scientific and some ideological. This group emphasizes the higher range of risk estimates and the lower range of benefit estimates. It is joined by a number of influential "public interest" groups — which actually are *interest* groups with an emphasis on their own view of the greater public good. Thus the environmentalist lobby has uncritically accepted as gospel the "worst-case scenarios" of danger to the public health and the environment. The minds of environmentalists were prepared by their earlier conclusion that nuclear science and technology are the tools of iniquitous social forces joined in an assault on the environment and the people. They view the expansion of knowledge as a worthwhile, long-term goal but one that can be postponed in a given political situation in favor of more pressing considerations. They see research as part of a social machine that can be stopped and started at the flick of a switch. To scientists the notion of postponing the quest for knowledge is a simplistic one that does not take into account the complexity and sensitivity of the requirements for successful research.

The press too is an interest group of consequence that finds grist for its own mill in emphasizing the more spectacular alternative scenarios and the clash of the more charismatic personalities. In ferreting out its version of the public interest the press features confrontation and walks quietly away from the duller details of the resolution. Thus the impression conveyed to the public, at least the fraction whose attention is engaged, is of a series of crucial battles. Where there are battles there must be armies, victors, and vanquished. Though it champions the right to know and serves to create an awareness of problems, the press cannot foster an optimum climate for formulating sound policy in a novel and technically complex area.

These individual interests have publicly defined the issues and the parameters for their solution in the public interest. The public interest, however, is an abstraction. Everyone shares in it but no individual interest truly represents it. Moreover, for routine issues the arena in which a decision is made never includes the total public. Political procedures are initiated or altered and choices are made within a more limited decision arena. Thus, the "system" moves ponderously along with little general public attention, its only "fine tuning" the result of clashes among groups with restricted interests. This is the way regulation of recombinant-DNA research is coming about — as a routine matter handled in a routine way. Yet in the background — and possibly being decided simultaneously — lie political issues that are far from routine.

Objective and Purpose

What are the deeper political issues surrounding regulation? For one thing, the sharp rise of both the cost and the consequence of incremental knowledge have fortified a growing feeling that the directions that research will take cannot be left to researchers alone. Potential biohazard called attention to recombinant DNA, and subsequently it reached partial public awareness. But that is not the whole problem. Recombinant DNA, as well as other advances of molecular genetics and cell biology, are opening wide new avenues to biological intervention. In turn, these avenues are possible paths to control of important aspects of the human future.

Recombinant DNA was not discovered at the behest of planners of the human future any more than was atomic fission. In truth, despite the many anxieties about it, there are no effective planners of the human future. Both atomic fission and molecular genetics, the products of fundamental physics and biology, clearly are important determinants of what is to come. Should our future depend only upon the *inner* workings of the scientific process, described earlier as driven largely by the self-propulsive "logic of science"? Will such an autonomous knowledge process always serve the public interest? Are there considerations other than the most effective way to gain new knowledge that should steer the human future?

This is the deep issue raised in the recombinant-DNA research debate — how the clearly successful but inner-directed search for knowledge can be accommodated to a larger sense of purpose in the greater community. Shall self-unfolding knowledge behave like blind fate to determine our future or shall our hopes and purposes contribute equally to our choices? Research choices are slowly being coupled more firmly to public decision making. Are the mechanisms by which this is happening appropriate and beneficial to both research and purpose? Or is it time to examine these mechanisms and redesign them for more conscious and deliberate interaction?

If these questions underlie the recombinant-DNA debate, the issue is not a routine one. The issue needs to engage more than the restricted number of interests so far involved. It needs more than a jerry-built regulatory agency or a knee-jerk response that could overshoot in one dimension through ignorance of other more important ones. To paraphrase a traditional American battle cry, regulation without full consideration is analogous to taxation without representation. If recombinant DNA is a bellwether issue, perhaps it is time to consider a more comprehensive approach to our system of husbandry of knowledge. For knowledge is providing us ever greater capability for intervention — in almost every sphere of action. The question is how best to keep these interventions within the orbit of objective and purpose.

Epilogue:
Agenda for
the Future

The first controlled DNA recombinations were performed five years ago. The NIH guidelines were promulgated three years later. The guidelines placed significant limitations on the conduct of important basic research but left many unanswered policy questions. At this writing, in the summer of 1978, it is not clear how this remaining policy agenda will be dealt with.

After extensive public hearings in both houses, statutory regulation of recombinant-DNA research appears to have been dropped by the Ninety-fifth Congress. Following a sharp clash between proponents of the two approaches to regulation — an independent public commission versus the executive authority of the Secretary of Health, Education and Welfare — congressional appetite for involvement declined sharply as expert estimates of possible hazard were scaled down and scientific resistance to regulation scaled up. The resulting stalemate leaves public policy on recombinant-DNA research incomplete and unstable.

The policy is incomplete because existing quasi-regulation applies only to federally supported research, and particularly because it fails to reach the expanding activity anticipated in the private, commercial sector. Policy is unstable for three reasons. First, the failure of Congress to adopt a firm federal stance invites a plethora of state and local initiatives targeted at major research centers in politically susceptible locales. Second, academic institutions that are primarily affected by existing regulations may not continue to treat them seriously if comparable commercial activities remain regulation-free. Third, a series of issues beyond biohazard affords legitimate toeholds for a continuing critical attack on recombinant-DNA research. Congressional equivocation has therefore not only failed to achieve political closure but it has left dangling ends that threaten to undercut interim measures already in place.

 Thus the political agenda clearly must be reviewed. First, the matter of risk and its assessment needs continuing attention. Some of the worst case scenarios are no longer seriously and widely entertained. With passing time their shock effect diminished and in the absence of any demonstrated untoward effect they lost persuasiveness. Yet legitimate concern about risk has not been totally laid to rest. The NIH Recombinant DNA Molecule Program Advisory Committee, in recently recommending considerable reduction of the stringency of the original guidelines, comments "... everything we have learned tends to diminish our estimate of the risk associated with recombinant DNA in *E. coli* K– 12. Nevertheless, the revised Guidelines continue to be deliberately restrictive, with the intent of erring on the side of caution."*

This reassurance with respect to possible hazard of *E. coli* K– 12 experimental systems under the guidelines is important, since this was a major focus of initial concern and K– 12 systems are still the major experimental material to date. However, the flexibility and wide potential applicability of recombinant-DNA techniques assuredly will bring many other experimental systems under consideration. There remains the urgent question of the means and mechanisms for effective risk assessment with respect to these systems. Experience with the K– 12 case indicates how laborious and time-consuming the process of assessment is. Other cases, for example plant systems for agricultural research, require very different expertise and orientation from that of the NIH. It is not clear that the current assignment of risk assessment to NIH alone fits either the institutes' capability or motivation. This concern is especially acute if the absence of statutory federal policy means that no new funds will be available to finance effective risk assessment.

 A second agenda item is assigning responsibility for monitoring the further advances of molecular genetic research so as to anticipate consequences and plan appropriate reactions. Enough has been said during the debate to make two things clear: (1) major practical consequences of already existing knowledge can be expected and additional consequential knowledge almost certainly will emerge within a decade; and (2) the public anxiety level is high, the mass media are alerted, and any even small untoward event is likely to be magnified in adverse impact.

*Federal Register, Vol. 42, No. 187, p. 49597, September 27, 1977.

Therefore, responsible scientific and public decision making of a high
order is called for. Greater unity of the two than has so far been displayed is essential.

A third major agenda item is the continuing effort to match public
expectation and aspiration with directions of molecular genetic research. The issue is, of course, not confined to recombinant DNA and
molecular genetics. Nonetheless, the tension is sharply delineated in
this area, because the special social and scientific aspects have created
what is almost a test case. The sensitivities surrounding the issue have
clearly been displayed in the polarization between those who would
ban the research entirely and those who insist that the research process
is sacrosanct against any form of external influence or control. Neither
of these extreme views is widely supported nor feasible in the real
world. Nonetheless, they represent conflicting orientations that have
been and will continue to be in contention. Bringing those who hold
these and intermediate views into accord is the role of the political
process, and it is necessary that we provide suitable arenas in which
such accommodation can be achieved. The history of the recombinant-DNA debate suggests that all parties have been dissatisfied
with existing arenas for the discussion and that the bitterness of the past
debate sprang in part from frustration with the available means for
expression and accommodation of views prior to decision making. No
progress can be made on this fundamental step in policy formation
unless a better mechanism is devised for organizing the process.

This matter of the debate arena emphasizes the fourth point on the
policy agenda. Consideration of policy governing recombinant-DNA research began in the involved research community, expanded to a
somewhat larger scientific and federal policy sphere, and then
exploded into a series of fragmented exchanges in the body politic.
When the political energy expended exceeded the congressional noise
level the legislative decision-making process was initiated. What followed seems to justify the contention that the issue as it then existed —
and by extension as it exists now — was not ready for definitive congressional action. Too little was known for certain, the situation was unprecedented and changing rapidly, and only a small and specialized fraction
of the public was aware of the issue and could even begin to understand
it. Nonetheless, Congress set about dealing with the matter by reaching
into its standard bag of tricks, both procedurally and in terms of available mechanisms. Meanwhile, the executive branch sought to fit the
issue into the existing bureaucracy, seemingly oblivious to the fact that
the issue cut sharply across the bureaucratic structure. This is especially
clear in the failure to recognize the important future implications for
the Departments of Commerce and Agriculture.

The issue, in fact, was not and is not ready for bureaucratic decision
making or for definitive congressional action. What is needed is what
the Ninety-fifth Congress gradually moved toward but did not quite
achieve. The indicated approach has three components: (1) establish-

ment of the federal presence in an area of profound national and international significance; (2) assurance of uniform applicability of policy to all recombinant-DNA research conducted within the United States; and (3) creation of an arena in which all issues can be examined and which gives full access to all interested parties and experts. My own view is that a presidential study commission, with all of its possible shortcomings, is the most appropriate arena for the kind of comprehensive assessment required. Had such a commission been established when the NIH guidelines were promulgated we might have been spared two years of bitter frustration and might now be almost ready for sound policy making. Fortunately, the reduced assessments of *current* risk make this course still timely and appropriate.

Such a commission would confront a specific case of one of the most critical issues involving science that must now be resolved through public policy: How to relate the increasing powers of biological science to the realization of human aspirations. We are nearing the end of a millennium that emerged painfully from the Dark Ages. The twentieth century has already brought a scientific crescendo of unprecedented magnitude. It has laid a platform for human choices that can make the next millennium either a final one or what ancient sages foresaw as the promised one.

It is not surprising, then, that anxious questions are now being raised about the power and potential of knowledge and its impact on the human future. Accumulating knowledge has emerged as a major and essential factor in guiding human directions. But what determines the directions of knowledge generation itself? How and by whom are the choices being made? Does research sometimes set in motion trends that neither researchers nor such social planners as we have intend? The human species has long confronted uncertainties in natural events that may lead, for example, to famine or plenty. Just as we become more able to resolve these uncertainties there arise new uncertainties about the consequences of our own inventions. Is the march of human knowledge and invention as inexorable and yet unpredictable in its consequences as flood, drought, and earthquake? Or is the knowledge process amenable to purposeful control? If it is, how can we institutionalize the necessary judgment to control it wisely?

These are the deep questions underlying the double image of the double helix, a current version of the Faustian dilemma. The classic dilemma seems to become more insistent even as seizing one horn or the other becomes less possible. Clearly, we must now not only live with uncertainty but also have the courage to exploit it. There was no certainty in the long evolutionary process that produced us. Rather, we arose through individually improbable events in concatenations of higher probability. Now we need more effective and deliberate ways to make bolder and still more successful combinations against uncertainty.

Whether we wish to regulate our own numbers, assure an adequate food supply, or step off into space, the crucial requirement is that purpose and knowledge guide action. In our origins we discovered

the power of this coupling. As our society has become more complex purpose, knowledge, and action have become, in some measure, institutionally separated. Those who expand our knowledge of the molecular basis of heredity cannot determine the purposes it will fulfill. Those who have responsibility to set social objectives are surprised by and do not fully comprehend so radical an increment to knowledge. And those whose role and satisfaction involve keeping things moving have no time or inclination to debate either the purposes or the implications of applying the new knowledge. Thus, recombinant DNA epitomizes the challenge to find new ways to rejoin purpose, knowledge, and action, the essential ingredients for human survival in the long struggle with uncertainty.

APPENDIX I

Summary Statement of the Asilomar Conference on Recombinant DNA Molecules, May 1975*

PAUL BERG†, DAVID BALTIMORE‡, SYDNEY BRENNER§, RICHARD O. ROBLIN III¶, AND MAXINE F. SINGER||

Organizing Committee for the International Conference on Recombinant DNA Molecules, Assembly of Life Sciences, National Research Council, National Academy of Sciences, Washington, D.C. 20418. †Chairman of the committee and Professor of Biochemistry, Department of Biochemistry, Stanford University Medical Center, Stanford, California; ‡American Cancer Society Professor of Microbiology, Center for Cancer Research, Massachusetts Institute of Technology, Cambridge, Mass.; §Member, Scientific Staff of the Medical Research Council of the United Kingdom, Cambridge, England; ¶Professor of Microbiology and Molecular Genetics, Harvard Medical School, and Assistant Bacteriologist, Infectious Disease Unit, Massachusetts General Hospital, Boston, Mass.; and || Head, Nucleic Acid Enzymology Section, Laboratory of Biochemistry, National Cancer Institute, National Institutes of Health, Bethesda, Maryland

I. INTRODUCTION AND GENERAL CONCLUSIONS

This meeting was organized to review scientific progress in research on recombinant DNA molecules and to discuss appropriate ways to deal with the potential biohazards of this work. Impressive scientific achievements have already been made in this field and these techniques have a remarkable potential for furthering our understanding of fundamental biochemical processes in pro- and eukaryotic cells. The use of recombinant DNA methodology promises to revolutionize the practice of molecular biology. Although there has as yet been no practical application of the new techniques, there is every reason to believe that they will have significant practical utility in the future.

Of particular concern to the participants at the meeting was the issue of whether the pause in certain aspects of research in this area, called for by the Committee on Recombinant DNA Molecules of the National Academy of Sciences, U.S.A. in the letter published in July, 1974** should end; and, if so, how the scientific work could be undertaken with minimal risks to workers in laboratories, to the public at large, and to the animal and plant species sharing our ecosystems.

The new techniques, which permit combination of genetic information from very different organisms, place us in an area of biology with many unknowns. Even in the present, more limited conduct of research in this field, the evaluation of

---

*Summary statement of the report submitted to the Assembly of Life Sciences of the National Academy of Sciences and approved by its Executive Committee on 20 May 1975.

Requests for reprints should be addressed to: Division of Medical Sciences, Assembly of Life Sciences, National Academy of Sciences, 2101 Constitution Avenue, N.W., Washington, D.C. 20418.

** Report of Committee on Recombinant DNA Molecules: "Potential Biohazards of Recombinant DNA Molecules," *Proc. Nat. Acad. Sci. USA* 71, 2593–2594, 1974.

potential biohazards has proved to be extremely difficult. It is this ignorance that has compelled us to conclude that it would be wise to exercise considerable caution in performing this research. Nevertheless, the participants at the Conference agreed that most of the work on construction of recombinant DNA molecules should proceed provided that appropriate safeguards, principally biological and physical barriers adequate to contain the newly created organisms, are employed. Moreover, the standards of protection should be greater at the beginning and modified as improvements in the methodology occur and assessments of the risks change. Furthermore, it was agreed that there are certain experiments in which the potential risks are of such a serious nature that they ought not to be done with presently available containment facilities. In the longer term, serious problems may arise in the large scale application of this methodology in industry, medicine, and agriculture. But it was also recognized that future research and experience may show that many of the potential biohazards are less serious and/or less probable than we now suspect.

II. PRINCIPLES GUIDING
THE RECOMMENDATIONS
AND CONCLUSIONS

Although our assessments of the risks involved with each of the various lines of research on recombinant DNA molecules may differ, few, if any, believe that this methodology is free from any risk. Reasonable principles for dealing with these potential risks are: (i) that containment be made an essential consideration in the experimental design and, (ii) that the effectiveness of the containment should match, as closely as possible, the estimated risk. Consequently, whatever scale of risks is agreed upon, there should be a commensurate scale of containment. Estimating the risks will be difficult and intuitive at first but this will improve as we acquire additional knowledge; at each stage we shall have to match the potential risk with an appropriate level of containment. Experiments requiring large scale operations would seem to be riskier than equivalent experiments done on a small scale and, therefore, require more stringent containment procedures. The use of cloning vehicles or vectors (plasmids, phages) and bacterial hosts with a restricted capacity to multiply outside of the laboratory would reduce the potential biohazard of a particular experiment. Thus, the ways in which potential biohazards and different levels of containment are matched may vary from time to time, particularly as the containment technology is improved. The means for assessing and balancing risks with appropriate levels of containment will need to be reexamined from time to time. Hopefully, through both formal and informal channels of information within and between the nations of the world, the way in which potential biohazards and levels of containment are matched would be consistent.

Containment of potentially biohazardous agents can be achieved in several ways. The most significant contribution to limiting the spread of the recombinant DNAs is the use of biological barriers. These barriers are of two types: (i) fastidious bacterial hosts unable to survive in natural environments, and (ii) nontransmissible and equally fastidious vectors (plasmids, bacteriophages, or other viruses) able to grow only in specified hosts. Physical containment, exemplified by the use of suitable hoods, or where applicable, limited access or negative pressure laboratories, provides an additional factor of safety. Particularly important is strict adherence to good microbiological practices which, to a large measure can limit the escape of organisms from the experimental situation, and thereby increase the safety of the operation. Consequently, education and training of all personnel involved in the experiments is essential to the effectiveness of all containment measures. In practice, these different means of containment will complement one another and documented substantial improvements in the ability to restrict the growth of bacterial hosts and vectors could permit modifications of the complementary physical containment requirements.

Stringent physical containment and rigorous laboratory procedures can reduce but not eliminate the possibility of spreading potentially hazardous agents. Therefore, investigators relying upon "disarmed" hosts and vectors for additional safety must rigorously test the effectiveness of these agents before accepting their validity as biological barriers.

III. RECOMMENDATIONS
FOR MATCHING TYPES OF
CONTAINMENT WITH
TYPES OF EXPERIMENTS

No classification of experiments as to risk and no set of containment procedures can anticipate all situations. Given our present uncertainties about the hazards, the parameters proposed here are broadly conceived and meant to provide provisional guidelines for investigators and agencies concerned with research on recombinant DNAs. However, each investigator bears a responsibility for determining whether, in his particular case, special circumstances warrant a higher level of containment than is suggested here.

A. TYPES OF CONTAINMENT

1. Minimal Risk   This type of containment is intended for experiments in which the biohazards may be accurately assessed and are expected to be minimal. Such containment can be achieved by following the operating procedures recommended for clinical microbiological laboratories. Essential features of such facilities are no drinking, eating, or smoking in the laboratory, wearing laboratory coats in the work area, the use of cotton-plugged pipettes or preferably mechanical pipetting devices, and prompt disinfection of contaminated materials.

2. Low Risk   This level of containment is appropriate for experiments which generate novel biotypes but where the available information indicates that the recombinant DNA cannot alter appreciably the ecological behavior of the recipient species, increase significantly its pathogenicity, or prevent effective treatment of any resulting infections. The key features of this containment (in addition to the minimal procedures mentioned above) are a prohibition on mouth pipetting, access limited to laboratory personnel, and the use of biological safety cabinets for procedures likely to produce aerosols (e.g., blending and sonication). Though existing vectors may be used in conjunction with low risk procedures, safer vectors and hosts should be adopted as they become available.

3. Moderate Risk   Such containment facilities are intended for experiments in which there is a probability of generating an agent with a significant potential for pathogenicity or ecological disruption. The principle features of this level of containment, in addition to those of the two preceding classes, are that transfer operations should be carried out in biological safety cabinets (e.g., laminar flow hoods), gloves should be worn during the handling of infectious materials, vacuum lines must be protected by filters, and negative pressure should be maintained in the limited access laboratories. Moreover, experiments posing a moderate risk must be done only with vectors and hosts that have an appreciably impaired capacity to multiply outside of the laboratory.

4. High Risk   This level of containment is intended for experiments in which the potential for ecological disruption or pathogenicity of the modified organism could be severe and thereby pose a serious biohazard to laboratory personnel or the public. The main features of this type of facility, which was designed to contain highly infectious microbiological agents, are its isolation from other areas by air locks, a negative pressure environment, a requirement for clothing changes and showers for entering personnel, and laboratories fitted with treatment systems to inactivate or remove biological agents that may be contaminants in exhaust air and liquid and solid wastes. All persons occupying these areas should wear protective laboratory clothing and shower at each exit from the containment facility. The handling of agents should be confined to biological safety cabinets in which the exhaust air is incinerated or passed through Hepa filters. High risk containment includes, in addition to the physical and procedural features described above, the use of rigorously tested vectors and hosts whose growth can be confined to the laboratory.

B. TYPES OF EXPERIMENTS

Accurate estimates of the risks associated with different types of experiments are difficult to obtain because of our ignorance of the probability that the anticipated dangers will manifest themselves. Nevertheless, experiments involving the construction and propagation of recombinant DNA molecules using DNAs from (*i*) prokaryotes, bacteriophages, and other plasmids, (*ii*) animal viruses, and (*iii*) eukaryotes have been characterized as minimal, low, moderate, and high risks to guide investigators in their choice of the appropriate containment. These designations should be viewed as interim assignments which will need to be revised upward or downward in the light of future experience.

The recombinant DNA molecules themselves, as distinct from cells carrying them, may be infectious to bacteria or higher organisms. DNA preparations from these experiments, particularly in large quantities, should be chemically inactivated before disposal.

1. Prokaryotes, Bacteriophages, and Bacterial Plasmids   Where the construction of recombinant DNA molecules and their propagation involves prokaryotic agents that are known to exchange genetic information naturally, the experiments can be performed in minimal risk containment facilities. Where such experiments pose a potential hazard, more stringent containment may be warranted.

Experiments involving the creation and propagation of recombinant DNA molecules from DNAs of species that ordinarily do not exchange genetic information, generate novel biotypes. Because such experiments may pose biohazards greater than those associated with the original organisms, they should be performed, at least, in low risk containment facilities. If the experiments involve either pathogenic organisms or genetic determinants that may increase the pathogenicity of the recipient species, or if the transferred DNA can confer upon the recipient organisms new metabolic activities not native to these species and thereby modify its relationship with the environment, then moderate or high risk containment should be used.

Experiments extending the range of resistance of established human pathogens to therapeutically useful antibiotics or disinfectants should be undertaken only under moderate or high risk containment, depending upon the virulence of the organism involved.

2. Animal Viruses Experiments involving linkage of viral genomes or genome segments to prokaryotic vectors and their propagation in prokaryotic cells should be performed only with vector-host systems having demonstrably restricted growth capabilities outside the laboratory and with moderate risk containment facilities. Rigorously purified and characterized segments of non-oncogenic viral genomes or of the demonstrably non-transforming regions of oncogenic viral DNAs can be attached to presently existing vectors and propagated in moderate risk containment facilities; as safer vector-host systems become available such experiments may be performed in low risk facilities.

Experiments designed to introduce or propagate DNA from non-viral and other low risk agents in animal cells should use only low risk animal DNAs as vectors (e.g., viral, mitochondrial) and manipulations should be confined to moderate risk containment facilities.

3. Eukaryotes The risks associated with joining random fragments of eukaryote DNA to prokaryotic DNA vectors and the propagation of these recombinant DNAs in prokaryotic hosts are the most difficult to assess.

A priori, the DNA from warm-blooded vertebrates is more likely to contain cryptic viral genomes potentially pathogenic for man than is the DNA from other eukaryotes. Consequently, attempts to clone segments of DNA from such animal and particularly primate genomes should be performed only with vector-host systems having demonstrably restricted growth capabilities outside the laboratory and in a moderate risk containment facility. Until cloned segments of warm-blooded vertebrate DNA are completely characterized, they should continue to be maintained in the most restricted vector-host system in moderate risk containment laboratories; when such cloned segments are characterized, they may be propagated as suggested above for purified segments of virus genomes.

Unless the organism makes a product known to be dangerous (e.g., toxin, virus), recombinant

DNAs from cold-blooded vertebrates and all other lower eukaryotes can be constructed and propagated with the safest vector-host system available in low risk containment facilities.

Purified DNA from any source that performs known functions and can be judged to be nontoxic, may be cloned with currently available vectors in low risk containment facilities. (Toxic here includes potentially oncogenic products or substances that might perturb normal metabolism if produced in an animal or plant by a resident microorganism.)

4. Experiments to be Deferred There are feasible experiments which present such serious dangers that their performance should not be undertaken at this time with the currently available vector-host systems and the presently available containment capability. These include the cloning of recombinant DNAs derived from highly pathogenic organisms (i.e., Class III, IV, and V etiologic agents as classified by the United States Department of Health, Education and Welfare), DNA containing toxic genes, and large scale experiments (more than 10 liters of culture) using recombinant DNAs that are able to make products potentially harmful to man, animals, or plants.

IV. IMPLEMENTATION

In many countries steps are already being taken by national bodies to formulate codes of practice for the conduct of experiments with known or potential biohazard.[††,‡‡] Until these are established, we urge individual scientists to use the proposals in this document as a guide. In addition, there are some recommendations which could be immediately and directly implemented by the scientific community.

A. DEVELOPMENT OF SAFER VECTORS AND HOSTS

An important and encouraging accomplishment of the meeting was the realization that special bacteria and vectors which have a restricted capacity to multiply outside the laboratory can be constructed genetically, and that the use of these organisms could enhance the safety of recombinant

---

[††] Advisory Board for the Research Councils, "Report of the Working Party on the Experimental Manipulation of the Genetic Composition of Micro-Organisms. Presented to Parliament by the Secretary of State for Education and Science by Command of Her Majesty, January 1975." London: Her Majesty's Stationery Office, 1975, 23pp.
[‡‡] National Institutes of Health Recombinant DNA Molecule Program Advisory Committee.

DNA experiments by many orders of magnitude. Experiments along these lines are presently in progress and in the near future, variants of Y bacteriophage, non-transmissible plasmids, and special strains of *Escherichia coli* will become available. All of these vectors could reduce the potential biohazards by very large factors and improve the methodology as well. Other vector-host systems, particularly modified strains of *Bacillus subtilis* and their relevant bacteriophages and plasmids, may also be useful for particular purposes. Quite possibly safe and suitable vectors may be found for eukaryotic hosts such as yeast and readily cultured plant and animal cells. There is likely to be a continuous development in this area and the participants at the meeting agreed that improved vector-host systems which reduce the biohazards of recombinant DNA research will be made freely available to all interested investigators.

B. LABORATORY PROCEDURES

It is the clear responsibility of the principal investigator to inform the staff of the laboratory of the potential hazards of such experiments before they are initiated. Free and open discussion is necessary so that each individual participating in the experiment fully understands the nature of the experiment and any risk that might be involved. All workers must be properly trained in the containment procedures that are designed to control the hazard, including emergency actions in the event of a hazard. It is also recommended that appropriate health surveillance of all personnel, including serological monitoring, be conducted periodically.

C. EDUCATION AND REASSESSMENT

Research in this area will develop very quickly and the methods will be applied to many different biological problems. At any given time it is impossible to foresee the entire range of all potential experiments and make judgments on them. Therefore, it is essential to undertake a continuing reassessment of the problems in the light of new scientific knowledge. This could be achieved by a series of annual workshops and meetings, some of which should be at the international level. There should also be courses to train individuals in the relevant methods since it is likely that the work will be taken up by laboratories which may not

have had extensive experience in this area. High priority should also be given to research that could improve and evaluate the containment effectiveness of new and existing vector-host systems.

V. NEW KNOWLEDGE

This document represents our first assessment of the potential biohazards in the light of current knowledge. However, little is known about the survival of laboratory strains of bacteria and bacteriophages in different ecological niches in the outside world. Even less is known about whether recombinant DNA molecules will enhance or depress the survival of their vectors and hosts in nature. These questions are fundamental to the testing of any new organism that may be constructed. Research in this area needs to be undertaken and should be given high priority. In general, however, molecular biologists who may construct DNA recombinant molecules do not undertake these experiments and it will be necessary to facilitate collaborative research between them and groups skilled in the study of bacterial infection or ecological microbiology. Work should also be undertaken which would enable us to monitor the escape or dissemination of cloning vehicles and their hosts.

Nothing is known about the potential infectivity in higher organisms of phages or bacteria containing segments of eukaryotic DNA and very little about the infectivity of the DNA molecules themselves. Genetic transformation of bacteria does occur in animals, suggesting that recombinant DNA molecules can retain their biological potency in this environment. There are many questions in this area, the answers to which are essential for our assessment of the biohazards of experiments with recombinant DNA molecules. It will be necessary to ensure that this work will be planned and carried out; and it will be particularly important to have this information before large scale applications of the use of recombinant DNA molecules is attempted.

The work of the committee was assisted by the National Academy of Sciences-National Research Council Staff: Artemis P. Simopoulos (Executive Secretary) and Elena O. Nightingale (Resident Fellow), Division of Medical Sciences, Assembly of Life Sciences, and suppoted by the National Institutes of Health (Contract NO1-OI)-5-2103) and the National Science Foundation (Grant GBMS75-05293).

Decision of the Director, National Institutes of Health, to Release Guidelines for Research on Recombinant DNA Molecules, June 1976

INTRODUCTION

Today, with the concurrence of the Secretary of Health, Education, and Welfare and the Assistant Secretary for Health, I am releasing guidelines that will govern the conduct of NIH-supported research on recombinant DNA molecules (molecules resulting from the recombination in cell-free systems of segments of deoxyribonucleic acid, the material that determines the hereditary characteristics of all known cells). These guidelines establish carefully controlled conditions for the conduct of experiments involving the insertion of such recombinant genes into organisms, such as bacteria. The chronology leading to the present guidelines and the decision to release them are outlined in this introduction.

In addition to developing these guidelines, NIH has undertaken an environmental impact assessment of these guidelines for recombinant DNA research in accordance with the National Environmental Policy Act of 1969 (NEPA). The guidelines are being released prior to completion of this assessment. They will replace the current Asilomar guidelines, discussed below, which in many instances allow research to proceed under

less strict conditions. Because the NIH guidelines will afford a greater degree of scrutiny and protection, they are being released today, and will be effective while the environmental impact assessment is under way.

Recombinant DNA research brings to the fore certain problems in assessing the potential impact of basic science on society as a whole, including the manner of providing public participation in those assessments. The field of research involved is a rapidly moving one, at the leading edge of biological science. The experiments are extremely technical and complex. Molecular biologists active in this research have means of keeping informed, but even they may fail to keep abreast of the newest developments. It is not surprising that scientists in other fields and the general public have difficulty in understanding advances in recombinant DNA research. Yet public awareness and understanding of this line of investigation is vital.

It was the scientists engaged in recombinant DNA research who called for a moratorium on certain kinds of experiments in order to assess the risks and devise appropriate guidelines. The capability to perform DNA recombinations, and the potential hazards, had become apparent at the Gordon Research Conference on Nucleic Acids in July 1973. Those in attendance voted to send an open letter to Dr. Philip Handler, President of the National Academy of Sciences, and to Dr. John R.

Federal Register, Vol. 41, No. 131, July 1976, pp. 27902–27911.

Hogness, President of the Institute of Medicine, NAS. The letter, appearing in *Science 181,* 1114, (1973), suggested "that the Academies [*sic*] establish a study committee to consider this problem and to recommend specific actions or guidelines, should that seem appropriate."

In response, NAS formed a committee, and its members published another letter in *Science 185,* 303, (1974). Entitled "Potential Biohazards of Recombinant DNA Molecules," the letter proposed:

"First, and most important, that until the potential hazards of such recombinant DNA molecules have been better evaluated or until adequate methods are developed for preventing their spread, scientists throughout the world join with the members of this committee in voluntarily deferring ... [certain] experiments....

"Second, plans to link fragments of animal DNAs to bacterial plasmid DNA or bacteriophage DNA should be carefully weighted...

"Third, the Director of the National Institutes of Health is requested to give immediate consideration to establishing an advisory committee charged with (i) overseeing an experimental program to evaluate the potential biological and ecological hazards of the above types of recombinant DNA molecules; (ii) developing procedures which will minimize the spread of such molecules within human and other populations; and (iii) devising guidelines to be followed by investigators working with potentially hazardous recombinant DNA molecules.

"Fourth, an international meeting of involved scientists from all over the world should be convened early in the coming year to review scientific progress in this area and to further discuss appropriate ways to deal with the potential biohazards of recombinant DNA molecules."

On October 7, 1974, the NIH Recombinant DNA Molecule Program Advisory Committee (hereafter "Recombinant Advisory Committee") was established to advise the Secretary, HEW, the Assistant Secretary for Health, and the Director, NIH, "concerning a program for developing procedures which will minimize the spread of such molecules within human and other populations, and for devising guidelines to be followed by investigators working with potentially hazardous recombinants."

The international meeting proposed in the *Science* article (*185,* 303, 1974) was held in February 1975 at the Asilomar Conference Center, Pacific Grove, California. It was sponsored by the National Academy of Sciences and supported by the National Institutes of Health and the National Science Foundation. One hundred and fifty people attended, including 52 foreign scientists from 15 countries, 16 representatives of the press, and 4 attorneys.

The conference reviewed progress in research on recombinant DNA molecules and discussed ways to deal with the potential biohazards of the work. Participants felt that experiments on construction of recombinant DNA molecules should proceed, provided that appropriate biological and physical containment is utilized. The conference made recommendations for matching levels of containment with levels of possible hazard for various types of experiments. Certain experiments were judged to pose such serious potential dangers that the conference recommended against their being conducted at the present time.

A report on the conference was submitted to the Assembly of Life Sciences, National Research Council, NAS, and approved by its Executive Committee on May 20, 1975. A summary statement of the report was published in *Science 188,* 991 (1975), *Nature 225,* 442, (1975), and the *Proceedings of the National Academy of Sciences 72,* 1981, (1975). The report noted that "in many countries steps are already being taken by national bodies to formulate codes of practice for the conduct of experiments with known or potential biohazard. Until these are established, we urge individual scientists to use the proposals in this document as a guide."

The NIH Recombinant Advisory Committee held its first meeting in San Francisco immediately after the Asilomar conference. It proposed that NIH use the recommendations of the Asilomar conference as guidelines for research until the committee had an opportunity to elaborate more specific guidelines, and that NIH establish a newsletter for informal distribution of information. NIH accepted these recommendations.

At the second meeting, held on May 12–13, 1975, in Bethesda, Maryland, the committee received a report on biohazard-containment facilities in the United States and reviewed a proposed NIH contract program for the construction and testing of microorganisms that would have very limited ability to survive in natural environments and would thereby limit the potential hazards. A subcommittee chaired by Dr. David Hogness was appointed to draft guidelines for research involving recombinant DNA molecules, to be discussed at the next meeting.

The NIH committee, beginning with the draft guidelines prepared by the Hogness subcommittee, prepared proposed guidelines for research with recombinant DNA molecules at its third meeting, held on July 18–19, 1975, in Woods Hole, Massachusetts.

Following this meeting, many letters were received which were critical of the guidelines. The majority of critics felt that they were too lax, others that they were too strict. All letters were reviewed by the committee, and a new subcommittee, chaired by Dr. Elizabeth Kutter, was appointed to revise the guidelines.

A fourth committee meeting was held on December 4– 5, 1975, in La Jolla, California. For this meeting a "variorum edition" had been prepared, comparing line-for-line the Hogness, Woods Hole, and Kutter guidelines. The committee reviewed these, voting item-by-item for their preference among the three variations and, in many cases, adding new material. The result was the "Proposed Guidelines for Research Involving Recombinant DNA Molecules," which were referred to the Director, NIH, for a final decision in December 1975.

As Director of the National Institutes of Health, I called a special meeting of the Advisory Committee to the Director to review these proposed guidelines. The meeting was held at NIH, Bethesda, on February 9– 10, 1976. The Advisory Committee is charged to advise the Director, NIH, on matters relating to the broad setting—scientific, technological, and socioeconomic—in which the continuing development of the biomedical sciences, education for the health professions, and biomedical communications must take place, and to advise on their implications for NIH policy, program development, resource allocation, and administration. The members of the committee are knowledgeable in the fields of basic and clinical biomedical sciences, the social sciences, physical sciences, research, education, and communications. In addition to current members of the committee, I invited a number of former committee members as well as other scientific and public representatives to participate in the special February session.

The purpose of the meeting was to seek the committee's advice on the guidelines proposed by the Recombinant Advisory Committee. The Advisory Committee to the Director was asked to determine whether, in their judgment, the guidelines balanced scientific responsibility to the public with scientific freedom to pursue new knowledge.

Public responsibility weighs heavily in this genetic research area. The scientific community must have the public's confidence that the goals of this profoundly important research accord respect to important ethical, legal, and social values of our society. A key element in achieving and maintaining this public trust is for the scientific community to ensure an openness and candor in its proceedings. The meetings of the Director's Advisory Committee, the Asilomar group, and the Recombinant Advisory Committee have reflected the intent of science to be an open community in considering the conduct of recombinant DNA experiments. At the Director's Advisory Committee meeting, there was ample opportunity for comment and an airing of the issues, not only by the committee members but by public witnesses as well. All major points of view were broadly represented.

I have been reviewing the guidelines in light of the comments and suggestions made by participants at that meeting, as well as the written comments received afterward. As part of that review I asked the Recombinant Advisory Committee to consider at their meeting of April 1– 2, 1976, a number of selected issues raised by the commentators. I have taken those issues and the response of the Recombinant Advisory Committee into account in arriving at my decision on the guidelines. An analysis of the issues and the basis for my decision follow.

I. GENERAL POLICY CONSIDERATIONS

A word of explanation might be interjected at this point as to the nature of the studies in question. Within the past decade, enzymes capable of breaking DNA strands at specific sites and of coupling the broken fragments in new combinations were discovered, thus making possible the insertion of foreign genes into viruses or certain cell particles (plasmids). These, in turn, can be used as vectors to introduce the foreign genes into bacteria or into cells of plants or animals in test tubes. Thus transplanted, the genes may impart their hereditary properties to new hosts. These cells can be isolated and cloned—that is, bred into a genetically homogeneous culture. In general, there are two potential uses for the clones so produced: as a tool for studying the transferred genes, and as a new useful agent, say for the production of a scarce hormone.

Recombinant DNA research offers great promise, particularly for improving the understanding and possibly the treatment of various diseases. There is also a potential risk—that microorganisms with transplanted genes may prove hazardous to man or other forms of life. Thus special provisions are necessary for their containment.

All commentators acknowledged the exemplary responsibility of the scientific community in dealing publicly with the potential risks in DNA recombinant research and in calling for a self-imposed moratorium on certain experiments in order to assess potential hazards and devise appropriate guidelines. Most commentators agreed that the process leading to the formulation of the proposed guidelines was a most responsible and responsive one. Suggestions by the commentators on broad policy considerations are presented below. They relate to the science policy aspects of the guidelines, the implementation of the guidelines for NIH grantees and contractors, and the scope and impact of the guidelines nationally and internationally.

A. SCIENCE POLICY CONSIDERATIONS

Commentators were divided on how best to steer a course between stifling research through excessive regulation and allowing it to continue with sufficient controls. Several emphasized that the public must have assurance that the controls afford adequate protection against potential hazards. In the views of these commentators, the burden is on the scientific community to show that the danger is minimal and that the benefits are substantial and far outweigh the risks.

Opinion differed on whether the proposed guidelines were an appropriate response to the potential benefits and hazards. Several found the guidelines to so exaggerate safety procedures that inquiry would be unnecessarily retarded, while others found the guidelines weighted toward promoting research. The issue was how to strike a reasonable balance, in fact, a proper policy "bias" — between concerns to "go slow" and those to progress rapidly.

There was strong disagreement about the nature and level of the possible hazards of recombinant DNA research. Several commentators believed that the hazards posed were unique. In their view, the occurrence of an accident or the escape of a vector could initiate an irreversible process, with a potential for creating problems many times greater than those arising from the multitude of genetic recombinations that occur spontaneously in nature. These commentators stress the moral obligation on the part of the scientific community to do no harm.

Other commentators, however, found the guidelines to be adequate to the hazards posed. In their view, the guidelines struck an appropriate balance so that research could proceed cautiously. Still other commentators found the guidelines too onerous and restrictive in light of the potential benefits of this research for medicine, agriculture, and industry. Some felt that the guidelines are perhaps more stringent than necessary given the available evidence on the likelihood of hazards, but supported them as a compromise that would best serve the scientific community and the public at large. Many commentators urged that the guidelines be adopted as soon as possible to afford more specific direction to this research area.

I understand and appreciate the concerns of those who urge that this research proceed because of the benefits and of those who urge caution because of potential hazards. The guidelines issued today allow the research to go forward in a manner responsive and appropriate to hazards that may be realized in the future. The object of these guidelines is to ensure that experimental DNA recombination will have no ill effects on those engaged in the work, on the general public, or on the environment. The essence of their construction is subdivision of potential experiments by class, decision as to which experiments should be permitted at present, and assignment to these of certain procedures for containment of recombinant organisms.

Containment is defined as physical and biological. Physical containment involves the isolation of the research by procedures which have evolved over many years of experience in laboratories studying infectious microorganisms. P1 containment — the first *physical* containment level — is that used in most routine bacteriology laboratories. P2 and P3 afford increasing isolation of the research from the environment P4 represents the most extreme measures used for containing virulent pathogens, and permits no escape of contaminated air, wastes, or untreated materials. *Biological* containment is the use of vectors or hosts that are crippled by mutation so that the recombinant DNA is incapable of surviving under natural conditions.

The experiments now permitted under the guidelines involve no known additional hazard to the workers or the environment beyond the relatively low risk known to be associated with the source materials. The additional hazards are speculative and therefore not quantifiable. In a real sense they are considerably less certain than are the benefits now clearly derivable from the projected research.

For example, the ability to produce, through "molecular cloning," relatively large amounts of pure DNA from the chromosomes of any living organism will have a profound effect in many areas of biology. No other procedure, not even chemical synthesis, can provide pure material corresponding to particular genes. DNA "probes," prepared from the clones will yield precise evidence on the presence or absence, the organization, and the expression of genes in health and disease.

Potential medical advances were outlined by scientists active in this research area who were present at the meeting of the Director's Advisory Committee. Of enormous importance, for example, is the opportunity to explore the malfunctioning of cells in complicated diseases. Our ability to understand a variety of hereditary defects may be significantly enhanced, with amelioration of their expression a real possibility. There is the potential to elucidate mechanisms in certain cancers, particularly those that might be caused by viruses.

Instead of mere propagation of foreign DNA, the expression of the genes of one organism by the cell machinery of another may alter the new host and open opportunities for manipulating the biological properties of cells. In certain prokaryotes (organisms with a poorly developed nucleus, like bacteria), this exchange of genetic information occurs in nature. Such exchange explains, for instance, an important mechanism for the changing and spreading of resistance to antiobiotics in bacteria. Beneficial effects of this mechanism might be the production of medically important compounds for the treatment and control of disease. Examples frequently cited are the production of insulin, growth hormone, specific antibodies, and clotting factors absent in victims of hemophilia.

Aside from the potential medical benefits, a whole host of other applications in science and technology have been envisioned. Examples are the large-scale production of enzymes for industrial use and the development of bacteria that could ingest and destroy oil spills in the sea. Potential benefits in agriculture include the enhancement of nitrogen fixation in certain plants, permitting increased food production.

While the projected research offers the possibility of many benefits, it must proceed only with assurance that potential hazards can be controlled or prevented. Some commentators are concerned that nature may maintain a barrier to the exchange of DNA between prokaryotes and eukaryotes (higher organisms, with a well-formed nucleus)—a barrier that can now be crossed by experimentalists. They further argue that expression of the foreign DNA may alter the host in unpredictable and undesirable ways. Conceivable harm could result if the altered host has a competitive advantage that would foster its survival in some niche within the ecosystem. Other commentators believe that the endless experiments in recombination of DNA which nature has conducted since the beginning of life on the earth, and which have accounted in part for the evolution of species, have most likely involved exchange of DNA between widely disparate species. They argue that prokaryotes such as bacteria in the intestines of man do exchange DNA with this eukaryotic host and that the failure of the altered prokaryotes to be detected attests to a sharply limited capacity of such recombinants to survive. Thus nature, this argument runs, has already tested the probabilities of harmful recombination and any survivors of such are already in the ecosystem. The fact is that we do not know which of the above-stated propositions is correct.

The international scientific community, as exemplified by the Asilomar conference and the deliberations attendant upon preparation of the present guidelines, has indicated a desire to proceed with research in a conservative manner. And most of the considerable public commentary on the subject, while urging caution, has also favored proceeding. Three European groups have independently arrived at the opinion that recombinant DNA research should proceed with caution. These are the Working Party on Experimental Manipulation of the Genetic Composition of Micro-Organisms, whose "Ashby Report" was presented to Parliament in the United Kingdom by the Secretary of State for Education and Science in January 1975; the Advisory Committee on Medical Research of the World Health Organization, which issued a press release in July 1975; and the European Molecular Biology Organization Standing Committee on Recombinant DNA, meeting in February 1976.

There is no means for a flat proscription of such research throughout the world community of science. There is also no need to attempt it. It is likely that the evaluation engendered in the preparation and application of these guidelines will lead to beneficial review of some of the containment practices in other work that is not technically defined as recombinant DNA research.

Recombinant DNA research with which these guidelines are concerned involves microorganisms such as bacteria or viruses or cells of higher organisms growing in tissue culture. It is extremely important for the public to be aware that this research is not directed to altering of genes in humans although some of the techniques developed in this research may have relevance if this is attempted in the future.

NIH recognizes its responsibility to conduct and support research designed to determine the extent to which certain potentially harmful effects from recombinant DNA molecules may occur. Among these are experiments, to be conducted under maximum containment, that explore the capability of foreign genes to alter the character of host or vector, rendering it harmful, as through the production of toxic products.

Given the general desire that no rare and unexpected event arising from this research shall cause irreversible damage, it is obvious that merely to establish conservative rules of conduct for one group of scientists is not enough. The precautions must be uniformly and unanimously observed. Second, there must be full and timely exchange of experiences so that guidelines can be altered on the basis of new knowledge. The guidelines must also be implemented in a manner that protects all concerned—the scientific workers most likely to encounter unexpected hazards and all forms of life within our biosphere. The responsibility of the scientists involved is as inescapable and extreme as is their opportunity to beneficially enrich our understanding.

B. IMPLEMENTATION CONSIDERATIONS WITHIN THE NIH

All the commentators had suggestions concerning the structure and function of decision making as it relates to the principal investigator, the local biohazards committee, the peer review group, and the NIH Recombinant Advisory Committee. These comments and my response on the section of the guidelines relating to roles and responsibilities of investigators, their institutions, and the National Institutes of Health are presented below.

Of considerable concern to all commentators was the process by which NIH would proceed to implement the guidelines. The scientific community generally urged that there be no Federal regulations, while some of the public commentators recommend the regulatory process.

Many who opposed changing the proposed guidelines into Federal regulations expressed concern for flexibility and administrative efficiency, which could best be achieved, in their view, through voluntary compliance. Other commentators, however, believed it imperative to proceed toward regulation. In their view, the guidelines could be implemented for purposes of NIH funding and would govern the conduct of experiments until regulations were in effect. Another commentator who thought regulation would be harmful rather than helpful suggested that if there were to be regulations, they should be along lines similar to those that govern the sale, distribution, use, and disposal of radioisotopes.

The question of how best to proceed now that the guidelines have been released deserves careful attention. I share the concern of those who feel that the guidelines must remain flexible. It is especially important that there be opportunity to change them quickly, based on new information relating to scientific evidence, potential risks, or safety aspects of the research program.

The suggestions for regulation need further attention at this time. The process for regulation not only involves the Director of NIH, but also the Assistant Secretary for Health and the Secretary of Health, Education, and Welfare. These guidelines are being promulgated now in order to afford additional protection to all concerned. Consideration of their conversion to regulations can proceed with continuing review of their content and present and future implications. Meanwhile, the NIH shall continue to provide the opportunity for public comment and participation at least equivalent to that provided if steps towards regulations were to proceed immediately. The guidelines will be published in the FEDERAL REGISTER forthwith to allow for further public comment.

C. IMPLEMENTATION CONSIDERATIONS BEYOND THE PURVIEW OF NIH

Special concern has been expressed by many commentators regarding the application of the guidelines to research outside NIH by investigators other than its grantees or contractors. It has been urged that the guidelines be made applicable to recombinant DNA research conducted or supported by other agencies in HEW and by NSF, ERDA, DoD, and other governmental departments. Most commentators believe that these or similar guidelines should also govern research in the private sector, including industry, voluntary organizations, and foundations. Many feel that experiments conducted in colleges, universities, and even in high schools require some form of monitoring. And finally, all agree that in view of the potential hazards of recombinant DNA research to the biosphere, some form of international understanding on guidelines for the research is essential.

The committee, in the proposed guidelines, has suggested as one means of control that a description of the physical and biological containment procedures practiced in a research project be included in the publication of research results. In the scientific community this can be a powerful force for conformity, and we will undertake to present the recommendation to all appropriate journals. We are also prepared to take steps to disseminate the guidelines widely, and to arrange for a continual flow of information outward concerning the activities of the Recombinant Advisory Committee and the Advisory Committee to the Director, NIH, in the evolution of the guidelines and their implementation.

In response to these suggestions, I have already held a meeting with relevant HEW agencies and with representatives from other departments of the Federal Government. The purpose of the meeting was to exchange information on recombinant DNA research and to discuss the NIH guidelines. It served as an important beginning to address a common concern of these public institutions. A number of the representatives indicated that various departments might very well adopt the guidelines for research conducted both in-house and supported outside. Following up, I have begun preliminary discussions with the Assistant Secretary for Health and the Secretary of HEW, to determine possible methods to ensure adoption of the guidelines by all Federal agencies. Encouraged by these efforts, we held a meeting on June 2 with representatives of industry to provide them with full information about the guidelines and to help determine the present and future interests of industrial laboratories in this type of research. The meeting provided one of the first opportunities for industry representatives to convene for a discussion of this research area, and an industry committee under the auspices of the Pharmaceutical Manufacturers Association will be formed to review the guidelines for potential application to the drug industry. Further meetings will be scheduled with other groups that have an active interest in recombinant DNA research.

It is my hope that the guidelines will be voluntarily adopted and honored by all who support or conduct such research throughout the United States, and that at least very similar guidelines will obtain throughout the rest of the world. NIH places the highest priority on efforts to inform and to work with international organizations, such as the World Health Organization and the International Council of Scientific Unions, with a view to achieving a consensus on safety standards in this most important research area.

There has been considerable international cooperation and activity in the past, and I expect it to continue in the future. The aforementioned Ashby Report, presented to Parliament in January 1975, describes the advances in knowledge and possible benefits to society of the experiments involving recombinant DNA molecules, and attempts to assess the hazards in these techniques. The Asilomar meeting also had a number of international representatives, as mentioned previously. The European Molecular Biology Organization (EMBO) has been involved in considering guidelines for recombinant DNA research. They have closely followed the activities of NIH, and will thus be encouraged, I believe, to monitor their research with augmented cooperation and coordination. For example, EMBO recently announced plans for a voluntary registry of recombinant DNA research in Europe. Following this EMBO initiative, NIH shall similarly maintain a voluntary registry of investigators and institutions engaged in such research in the United States. Plans for establishing this registry are under way.

D. ENVIRONMENTAL POLICY
CONSIDERATIONS

A number of commentators urged NIH to consider preparing an environmental impact statement on recombinant DNA research activity. They evoked the possibility that organisms containing recombinant DNA molecules might escape and affect the environment in potentially harmful ways.

I am in full agreement that the potentially harmful effects of this research on the environment should be assessed. As discussed throughout this paper, the guidelines are premised on physical and biological containment to prevent the release or propagation of DNA recombinants outside the laboratory. Deliberate release of organisms into the environment is prohibited. In my view, the stipulated physical and biological containment ensures that this research will proceed with a high degree of safety and precaution. But I recognize the legitimate concern of those urging that an environmental impact assessment be done. In view of this concern and ensuing public debate, I have reviewed the appropriateness of such an assessment and have directed that one be undertaken.

The purpose of this assessment will be to review the environmental effects, if any, of research that may be conducted under the guidelines. The assessment will provide further opportunity for all concerned to address the potential benefits and hazards of this most important research activity. I expect a draft of the environmental impact statement should be completed by September 1 for comment by the scientific community, Federal and State agencies, and the general public.

It should be noted that the development of the guidelines was in large part tantamount to conducting an environmental impact assessment. For example, the objectives of recombinant DNA research, and alternate approaches to reach those objectives, have been considered. The potential hazards and risks have been analyzed. Alternative approaches have been thoroughly considered to maximize safety and minimize potential risk. And an elaborate review structure has been created to achieve these safety objectives. From a public policy viewpoint, however, the environmental impact assessment will be yet another review that will provide further opportunity for the public to participate and comment on the conduct of this research.

II. METHODS OF CONTAINMENT

Comments on the containment provisions of the proposed guidelines were directed to the definition of both physical and biological containment and to the safety and effectiveness of the prescribed levels. Several commentators found the concept of physical containment imprecise and too subject to the possibility for human error. Others questioned the concept of biological containment in terms of its safety and purported effectiveness in averting potential hazards. The commentators were divided on which method of containment would provide the most effective and safe system to avoid hazards. Several suggested that each of the physical containment levels be more fully explained.

W. Emmett Barkley, Ph.D., Director of the Office of Research Safety, National Cancer Institute, was asked to review the section on physical containment in light of these comments. Dr. Barkley convened a special committee of safety and health experts, who met to consider not only this section of the guidelines but also the section on the roles and responsibilities of researchers and their institutions. The committee thoroughly reviewed the section on physical containment and recommended a number of changes. The Recombinant Advisory Committee, meeting on April 1–2, 1976, reviewed the recommendations of the Barkley group. These are incorporated, with editorial revisions, in the final version of the guidelines.

The present section on physical containment is directly responsive to those commentators who asked for greater detail and explanation. Although different in detail, the four levels of containment approximate those given by the Center for Disease Control for human etiologic agents and by the National Cancer Institute for oncogenic viruses. For each of the proposed levels, optional items have been excluded, and only those items deemed absolutely necessary for safety are presented. Necessary facilities, practices, and equipment are specified. To give further guidance to investigators and their institutions, a supplement to the guidelines explains more fully safety practices appropriate to recombinant DNA research. And a new section has been added to ensure that shipment of recombinant DNA materials conforms, where appropriate, to the standards, prescribed by the U.S. Public Health Service, the Department of Transportation, and the Civil Aeronautics Board.

The section on physical containment is carefully designed to offer a constructive approach to meeting potential hazards for recombinant experiments at all levels of presumed risk. Certain commentators had suggested that the first level of physical containment (P1) be merged with the second level (P2). This suggestion, however, would tend to apply overly stringent standards for some experiments and might result in a lowering of standards necessary at the second level. I believe the level of control must be consistent with a reasonable estimate of the hazard; and the section on physical containment does provide this consistency. Accordingly, the first and second levels of physical containment remain as separate sections in the guidelines.

Because of the nature and operation of facilities required for experiments to be done at the fourth level of containment (P4), a provision has been included that the NIH shall review such facilities prior to funding them for recombinant DNA studies. The situation merits the special attention of experts who have maximum familiarity with the structure, operation, and potential problems of P4 installations. Several commentators advocated that NIH arrange for sharing of P4 facilities, both in the NIH intramural program and in institutions supported through NIH awards. In response to these suggestions, we are currently reviewing our facilities, including those at the Frederick Cancer Research Center (Fort Detrick), to determine how such a program can best be devised. It is most important that P4 facilities be made available to investigators. It should be noted that incidents of infection by even the most highly infectious and dangerous organisms are extremely infrequent at P4 facilities, and therefore the potential for hazard in certain complex experiments in recombinant DNA research is considerably reduced.

III. PROHIBITED EXPERIMENTS

1. Practically all commentators supported the present prohibition of certain experiments. There were suggestions for a clearer definition of the prohibition of certain experiments where increased antibiotic resistance may result. And it was urged by some that the prohibition be broadened to include experiments that result in resistance to any antibiotic, irrespective of its use in medicine or agriculture. Consideration of such a suggestion must take into account that antibiotic resistance occurs naturally among bacteria, and that resistance is a valuable marker in the study of microbial genetics in general, and recombinants in particular.

In view of these concerns, however, the Recombinant Advisory Committee was asked to reconsider carefully the prohibition and related sections concerning antibiotic resistance. The committee noted that the prohibition relating to drug resistance was intended to ban those experiments that could compromise drug use in controlling disease agents in veterinary as well as human medicine and this is now clearly stated.

In the draft guidelines there were two statements concerning resistance to drugs which related to experiments with *E. coli*. The statements appeared to allow experiments that would extend the range of resistance of this bacterium to therapeutically useful drugs and disinfectants, and thus seemed to be in conflict with the general prohibition on such research. There are numerous reports in the scientific literature indicating that *E. coli* can acquire resistance to all antibiotics known to act against it. Since *E. coli* acquires resistance naturally, the prohibition directed against increasing resistance does not apply. The ambiguous statements have been deleted from the present guidelines. On the other hand, new language has been inserted in the section dealing with other prokaryote species to set containment levels for permitted experiments.*

2. The Recombinant Advisory Committee was also asked to clarify whether the prohibition of use of DNA derived from pathogenic organisms (those classified as 3, 4, and 5 by the Center for Disease Control, USPHS) also included the DNA from any host infected with these organisms. The committee explained that this prohibition did extend to experiments with cells known to be so infected. To avoid misunderstanding, the prohibition as now worded includes such cells. In addition, the prohibitions have been extended to include moderate-risk oncogenic viruses, as defined

---

*Specifically, experiments that would extend resistance to therapeutically useful drugs must use P3 physical containment plus a host-vector comparable to EK1, or P2 containment plus a host-vector comparable to EK2.

by the National Cancer Institute, and cells known to be infected with them.

3. Two other issues relating to the section on prohibited experiments were raised by Roy Curtiss III, Ph.D., Professor, Department of Microbiology, University of Alabama School of Medicine, Birmingham, who is a member of the Recombinant Advisory Committee. Dr. Curtiss noted that for the class of experiments prohibited on the basis of production of highly toxic substances, only substances from microorganisms were cited as examples. He suggested that other examples be included, such as venoms from insects and snakes. The committee approved the suggestion and I concur.

In the proposed guidelines, release of organisms containing recombinant DNA molecules into the environment was prohibited unless a series of controlled tests had been done to leave no reasonable doubt of safety. Dr. Curtiss felt that the guidelines should provide greater specificity for testing and should include some form of review prior to release of the organism. I have decided that the guidelines should, for the present, prohibit any deliberate release of organisms containing recombinant DNA into the environment. With the present limited state of knowledge, it seems highly unlikely that there will be in the near future, any recombinant organism that is universally accepted as being beneficial to introduce into the environment. When the scientific evidence becomes available that the potential benefits of recombinant organisms, particularly for agriculture, are about to be realized, then the guidelines can be altered to meet the needs for release. It is most important that the potential environmental impact of the release be considered.

IV. PERMISSIBLE EXPERIMENTS:
E. COLI K–12 HOST-VECTOR SYSTEMS

The continued use of E. coli as a host has drawn considerable comment, including some suggestions that its use be prohibited presently or within a specified time limit. It should be stressed that the use of E. coli as detailed in the guidelines is limited to E. coli K–12, a strain that has been carried in the laboratory for decades and does not involve the use of any strain of E. coli that is freshly isolated from a natural source. E. coli K–12 does not usually colonize the normal bowel, even when given in large doses, and exhibits little if any multiplication while passing through the alimentary canal. For years it has been the subject of more intense investigation than any other single organism, and knowledge of its genetic makeup and recombinant behavior exceeds greatly that pertaining to any other organism. I believe that because of this experience, E. coli

K–12 will provide a host-vector system that is safer than other candidate microorganisms.

NIH recognizes the importance of supporting the development of alternative host-vector systems (such as B. subtilis, which has no ecological niche in man) and will encourage such development. It should be noted, however, that for each new host-vector system, the same questions of risk from altered properties attendant upon the presence of recombinant genes will apply as apply to E. coli. NIH does not believe it wise to set a time limit on replacement of E. coli systems by other organisms.

There were specific suggestions concerning the three levels of biological containment prescribed for use of E. coli K–12 host-vectors. Some commentators requested a more detailed explanation of the adequacy of protection for laboratory personnel with the first level of containment (EK1).* Sections of the guidelines dealing with physical containment and roles and responsibilities now specify the need for safety practices and accident plans.

For the second level of containment (EK2), it is required that a cloned DNA fragment be contained in a host-vector system that has no greater than a 10^{-8} probability of survival in a nonpermissive or natural environment. It was suggested that the selection of this level of biological containment and the appropriate tests for verification be more fully explained in the guidelines. The committee, in responding to a request for further examination of this point, reviewed at considerable length the testing for an EK2 system and recommended certain modifications. We have accepted the committee's new language that better explains testing of survival of a genetic marker carried on the vector, preferably on an inserted NDA fragment.

Possible tests to determine the level of biological containment afforded by these altered host-vector systems are outlined in this section. Because this is such a new area of scientific research and development, however, it is inappropriate to standardize such testing at the present time.

---

* The EK1 system presently consists of a battery of different vectors and of E. coli K–12 mutants, all of which afford a considerable degree of biological containment. The diversity of vectors and of host mutants in this battery has permitted a wide range of important scientific questions to be attacked. For example, the availability of different vectors with cleavage sites for different restriction endonucleases have increased the kind of DNA segments that can be cloned. By contrast, the first EK2 host-vector systems are only now being considered by the Recombinant Advisory Committee. While NIH is supporting the development of more EK2 host-vector systems, it is not expected that a battery equivalent to that available for the EK1 system will be certified by the Recombinant Advisory Committee in the near future.

Standards will gradually be set as more experience with EK2 host-vector systems is acquired. The committee, for example, during its April 1976 meetings gave its first approval to an EK2 host-vector system. What is necessary is that new and more effective tests be devised by investigators, and this effort is very likely to occur under the present guidelines. For example, one task recognized by the committee is to clarify how survival of the organism and the cloned DNA should be defined in terms of temperature, medium, and other variables.

It is also very important to note here that the stringent requirements set by the committee for EK2 biological containment jeopardize considerably the capacity of such crippled organisms to survive and replicate even under permissive laboratory conditions. More experience will be required to determine whether EK2 containment will permit some lines of important research to be followed.

Several commentators suggested that methods and procedures to confirm an EK system at the third level of containment (EK3) be more fully explained. The Recombinant Advisory Committee was asked to consider this suggestion. After considerable discussion the committee declined to define the procedures more fully at this time, because development of an EK3 system is still far enough in the future not to warrant specific testing procedures. Further, it is not clear what tests are best suited. The language, therefore, remains general. The committee, however, is aware of the concerns for a more completely defined system of testing, and has considered the possibility of organizing a symposium for purposes of designating tests. In my view, more fully developed protocols for testing EK3 systems are warranted, and it is necessary that guidelines here be more fully developed before the committee proceeds to certify such a system. In this regard the NIH is prepared through the National Institute of Allergy and Infectious Diseases to support contracts to accomplish this task. We will seek the advice and assistance of the committee to define the scope of necessary work.

These guidelines also include a statement that for the time being no EK2 or EK3 host-vector system will be considered *bona fide* until the Recombinant Advisory Committee has certified it. I share the concern of the commentators that new host-vector systems require the highest quality of scientific review and scrutiny. At this early stage of development, it is most important that the committee provide that scrutiny. Further, I believe that until more experience has been gained, the committee should encourage and the NIH support research that will independently confirm and augment the data on which certification of EK2 host-vector systems are based.

V. CLASSIFICATION OF EXPERIMENTS USING THE *E. COLI* K–12 CONTAINMENT SYSTEMS

The guidelines assign different levels of containment for experiments in which DNA from different sources is to be introduced into an *E. coli* K–12 host-vector system. The variation is based on both facts and assumptions. There are some prokaryotes (bacteria) which constantly exchange DNA with *E. coli*. Here it is assumed that experimental conditions beyond those obtained in careful, routine microbiology laboratories are superfluous, because any exchange experiments have undoubtedly been performed already in nature.

In every instance of artificial recombination, consideration must be given to the possibility that foreign DNA may be translated into protein (expressed), and also to the possibility that normally repressed genes of the host may be expressed and thus change, undesirably, the characteristics of the cell. It is assumed that the more similar the DNAs of donor and host, the greater the probability of expression of foreign DNA, or of possible derepression of host genes. In those cases where the donor exchanges DNA with *E. coli* in nature, it is unlikely that recombination experiments will create new genetic combinations. When prokaryote donors not known to exchange DNA with *E. coli* in nature are used, however, there is a greater potential for new genetic combinations to be formed and be expressed. Therefore, it is required that experiments involving prokaryotic DNA from a donor that is not known to exchange DNA with *E. coli* in nature be carried out at a higher level of containment. Recombination using prokaryotic DNA from an organism known to be highly pathogenic is prohibited.

There are only limited data available concerning the expression of DNA from higher forms of life (eukaryotes) in *E. coli* (or any other prokaryote). Therefore, the containment prescriptions for experiments inserting eukaryotic DNA into prokaryotes are based on risks having quite uncertain probabilities.

On the assumption that a prokaryote host might translate eukaryotic DNA, it is further presumed that the product of that foreign gene would be most harmful to man if it were an enzyme, hormone, or other protein that was similar (homologous) to proteins already produced by or active in man. An example is a bacterium that could produce insulin. Such a "rogue" bacterium could be of benefit if contained, a nuisance or possibly dangerous if capable of surviving in nature. This is one reason that the higher the phylogenetic order of the eukaryote, the higher the recommended containment, at least until the efficiency of expression of DNA from higher eukaryotes in prokaryotes can be determined.

There is a second, more concrete reason for scaling containment upward as the eukaryote host becomes similar to man. This is the concern that viruses capable of propagating in human tissue, and possibly causing diseases, can contaminate DNA, replicate in prokaryote hosts and infect the experimentalist. Such risks are greatest when total DNA from donor tissue is used in "shotgun" recombinant experiments; it diminishes to much lower levels when pure cloned DNA is used.

The commentators were clearly divided on the classification of containment criteria for different kinds of recombinant DNAs. Many commentators considered the guidelines too stringent and rigid. Others viewed the guidelines in certain instances as too permissive. And still others endorsed the guidelines as sensible and reasonable, affording the public an enormous degree of protection from the speculative risks. Several suggestions were made for the specific classes of experiments, and they follow:

1. Comment on the use of DNA from animals and plants in recombinant experiments varied widely. Some commentators suggested banning the use of DNA from primates, other mammals, and birds. Others suggested that higher levels of containment be used for all such experiments. Still others believed that the guidelines were too strict for experiments of this class. I have carefully reviewed the issues raised by the commentators and the responses of the committee to certain queries concerning use of animal and plant DNA in these experiments.

In my view, the classification for the use of DNA from primates, other mammals, and birds is appropriate to the potential hazards that might be posed. The physical and biological containment levels are very strict. For example, biological containment levels are at EK2 or EK3, and will effectively preclude experimentation until useful EK2 and EK3 systems are available. EK2 systems are still in the initial stages of development, and the first system was only certified at the most recent meeting of the Recombinant Advisory Committee. An EK3 host-vector system has yet to be tested, and its certification is far enough in the future to place a moratorium on those experiments requiring biological containment at an EK3 level. The physical containment levels of P3 or P4 themselves afford a very high degree of protection. I am satisfied that the guidelines demonstrate the caution and prudence that must govern the conduct of experiments in this category.

The guidelines allow reduced containment levels for primate DNA when it is derived from embryonic tissue or germ-line cells. This is based on evidence that embryonic material is less likely to contain viruses than is tissue from the adult. Obviously, the embryonic tissue must be free of adult tissue, and the present guidelines so indicate.

I have also carefully considered the special concerns arising from the use of DNA from cold-blooded vertebrates and other cold-blooded animals, because several commentators questioned the basis of lower physical and biological containment levels for DNA from these species. The Recombinant Advisory Committee has debated this extensively, and they were asked to do so once again in April.* The committee has now recommended high containment levels (P3+EK2) when the DNA is from a cold-blooded vertebrate known to produce a potent toxin. That recommendation is included in the present guidelines. Where no toxin is involved the committee supported lower containment levels. The guidelines specify P2+EK2 levels for such work. There was considerable discussion concerning the advisability of recommending lower containment (P2+EK1) when the DNA is isolated from embryonic tissue or germ-line cells from cold-blooded vertebrates. Those supporting lower containment levels argued that the justification for P2+EK2 was the possibility that cold-blooded vertebrates may carry viruses and that the distinction between adult and germ-cell tissue is real. Others argued that, contrary to the situation with primate DNA, viruses are not a central problem with cold-blooded vertebrates and therefore no distinction should be made on the basis of tissue origin. Finally, the committee recommended, on a divided vote (8 to 4), to adopt P2+EK1 when the cold-blooded vertebrate DNA is isolated from embryonic tissue or germ-line cells. Upon reviewing these considerations, I have decided to retain the containment levels for embryonic or germ-line DNA from cold-blooded vertebrates as recommended by the committee.

---

*A committee member, David S. Hogness, Ph.D., Professor, Department of Biochemistry, Stanford University, California, submitted a statement in support of lower containment levels based on current scientific evidence. That evidence is based on certain differences between cold- and warm-blooded vertebrates. One of the criteria used for the evaluation of the relative risk that might be encountered with different levels of shotgun experiment is the degree of sequence homology between the DNA of the given species and that of humans. This criterion is used to estimate the likelihood that segments of DNA from the given species might be integrated into the human genome by recombination; the greater the homology, the greater the likelihood of integration. Studies of sequence homologies indicate that there is a considerable degree of homology between human DNA and DNA from other primates, much less homology between primates and other mammals, and even lower but detectable homology between birds and primates. By contrast, no significant homologies between cold-blooded vertebrates and primates have been detected.

In April the committee also reviewed, at our request, the classification of experiments where DNA is derived from other cold-blooded animals or lower eukaryotes. Several commentators, for example, had been concerned about the fact that insects are known to carry agents pathogenic to man. In the committee review it was noted that viruses carried by insects and known to transmit disease to man are RNA rather than DNA viruses and do not reproduce via DNA copied from RNA. In order, however, to make the intent clearer, the guidelines have been rewritten for experiments of this class. New language is inserted to ensure that strict containment levels are employed when the DNA comes from known pathogens or species known to carry them. Further, to reduce the potential hazards, we have also included in the guidelines the requirement that any insect must be grown under laboratory conditions for at least 10 generations prior to its use as a DNA source.

2. As alluded to above, certain commentators expressed concern that when *E. coli* becomes the host of recombinant DNA from prokaryotes with which DNA is not usually exchanged, there is hazard of altered host characteristics resulting from translation of the DNA into functioning proteins. The committee was asked to review the guidelines and take into account this potential hazard. They agreed that the containment levels should be increased for this category of experiment, from P2+EK1 to either P2+EK2 or P3+EK1. That recommendation is included in the present guidelines.

Comments were made concerning that class of experiments in which the recombinant DNA, regardless of source, has been cloned. A clone is a population of cells derived from a single cell and therefore all the cells are presumed to be genetically identical. As outlined in the proposed guidelines, clones could be used at lower containment levels if they had been rigorously characterized and shown to be free of harmful genes. Several commentators inquired how the characterization was to be performed and the freedom from harmful genes demonstrated. Although the committee acknowledges that these terms are unavoidably vague, they do cite appropriate scientific methods to make relevant determinations. Again, this is a rapidly changing area and more clarity and precision can be expected with experience. Reduced containment requirements for this class of experiment are warranted because of the purified nature of clones. Further, the granting agency must approve the clone before containment conditions can be reduced, thus providing an additional element of review.

4. Another comment was related to the use of DNA from organelles (intracellular elements that contain special groups of genes for particular cell functions). Concern was expressed about the potential contamination of purified organelle DNA with DNA from viruses because of the similarity of their structures. The committee agrees, and the guidelines now specify a requirement, that the organelles be isolated prior to extracting DNA, as a further means of reducing the hazard of viral contamination.

5. Some commentators were troubled about the lowering of containment for that class of experiments involving recombinations with cell DNA segments purified by chemical or physical methods. They asked that procedures for determining the state of purification be more fully detailed and that the Recombinant Advisory Committee certify the purity. There are, however, appropriate techniques, such as gel electrophoresis, with which a purity of 99 percent by mass can be achieved and ascertained. There is no way for the committee to certify these results beyond repeating the experiments themselves. These techniques are well documented and described in the literature. I do not believe it is necessary or feasible for the committee to review each procedure for purification of DNA.

6. Comments were made concerning the use of DNA derived from animal viruses. It was urged that containment levels for this class of experiment be increased. On the basis of my review, I find the containment conditions appropriate to the potential hazard posed. As defined in the guidelines, experiments are to be done at very strict levels of containment and these can be lowered only when the cloned DNA recombinants have been shown to be free of possibly harmful genes by suitable biochemical and biological tests. This also pertains to DNA that is copied from RNA viruses. In no instance are the guidelines more lenient, and in most instances they are more stringent than conditions obtaining in many laboratories where such viruses are studied in non-DNA-recombinant experiments.

VI. CLASSIFICATION OF EXPERIMENTS USING CONTAINMENT SYSTEMS OTHER THAN *E. COLI* K–12

1. No issue with regard to these guidelines raised more comment than the use of animal viruses as vectors. Of special concern to many commentators was the use of the simian (monkey) virus 40 (hereafter "SV40"). Some suggested a complete ban on the use of this virus; others urged its retention as a vector. SV40 is not known to produce any disease in man, although it can be grown in human cells and on very rare occasions has been isolated from humans. Many humans have received SV40 virus inadvertently in vaccines prepared from virus grown in monkey kidney-cell

cultures. An intensive search has been made and is continuing for evidence that SV40 might cause cancer or be otherwise pathogenic for man. At present, it is my view that the extensive knowledge we have of SV40 virus provides us with sufficient sophistication to ensure its safe handling under the conditions developed for its use in the guidelines.

I believe work with SV40 should continue under the cost careful conditions, but I do recognize and appreciate the concerns expressed over its possible harmful effects in humans. In light of these concerns, I asked the Recombinant Advisory Committee to review this section of the guidelines. The committee reconsidered the containment conditions for this class of experiments and judged them appropriate to meet the potential hazards.*

This class of experiments will proceed under the most careful and stringent conditions. Work with SV40 virus will be done at the maximum level of physical containment (P4). The extraordinary precautions required in a P4 facility lessen the likelihood of a potential hazard from this work. Only defective SV40 virus will be used as vector; that is, the SV40 virus particles that carry the foreign DNA cannot multiply by themselves. When a number of strict conditions are met, this work will be permitted to go on at the third level of containment (P3), which in itself requires care and precision. It should be noted that SV40 virus and its DNA can be efficiently disinfected by Clorox and autoclaving. These are customary procedures for disinfecting glassware and other items used in SV40 animal-cell work.

Some commentators suggested that the containment criteria for experiments using polyoma virus as the vector be strengthened. There is no evidence that polyoma infects humans or replicates to any significant extent in human cells. It holds promise as a vector, as is more fully documented in an appendix to these guidelines.

2. Several commentators found the guidelines inadequate regarding experiments with plant host-vector systems. Because NIH shared these concerns, a group with extensive experience with plants was appointed to review this section. The group met concurrently with the Recombinant Advisory Committee in April 1976 and made several modifications. The suggested revisions were acceptable to the full committee, and we have included them in the guidelines.

*One member dissented from this position. During the discussion, additional language was recommended (and adopted) to ensure that the defective SV40-virus/helper-virus system, with its inserted non-SV40 DNA segment, does not replicate in human cells with significantly more efficiency than does SV40.

The modifications are responsive to the stated concerns of the commentators. A description of greenhouse facilities is given, and physical containment conditions have been modified to take into account operations with whole plants. On the whole, the respective portions of the guidelines relating to plants are more fully explained and the intent is clarified.

I have also accepted the recommendation of the subcommittee to lower the biological containment level from EK2 to EK1 for experiments in which the DNA from plants is used in conjunction with the *E. coli* K– 12 host-vector system, thereby setting containment in this instance at the same level required for experiments with lower-eukaryote DNA.

VII. ROLES AND RESPONSIBILITIES

1. Most commentators had suggestions for the section on the roles and responsibilities of investigators, their local institutions, and NIH. Commentators generally urged openness, candor, and public participation in the process, emphasizing shared responsibility and accountability from the local to the national level. We reviewed that section of the guidelines in light of these comments and have asked the Recombinant Advisory Committee to review certain issues.

It is clear that much of the success of the guidelines will lie in the wisdom with which they are implemented. Because of the importance of this section, especially in terms of safety programs and plans, we have carefully weighed the comments and suggestions made in this regard. NIH has a special responsibility to take a leading role in ensuring that safety programs are part of all recombinant DNA research. Dr. Barkley and a specially convened committee were asked to provide greater detail for safety, accident, and training plans for this section of the guidelines. Based on their recommendations, the section has been extensively rewritten to clarify the respective responsibilities of the principal investigator, the institution (including the institutional biohazards committee), the NIH initial review group (study section), the NIH Recombinant DNA Molecule Program Advisory Committee, and NIH staff.

This section has a definitive administrative framework for assuring that safety is an essential and integrated component of research involving recombinant DNA molecules. The guidelines require investigators to institute, monitor, and evaluate containment and safety practices and procedures. Before research is done, the investigator must have safety and accident plans in place and training exercises for the staff well under way.

Some commentators suggested that the investigator be required to obtain informed consent of laboratory personnel prior to their participation. Rather than rely explicitly on an informed consent document, the guidelines now make the investigator responsible for advising his program and support staff as to the nature and assessment of the real and potential biohazards. He must explain and provide for any advised or requested precautionary medical policies, vaccinations, or serum collections. Further, an appendix to the guidelines includes detailed explanations for dealing with accidents, as well as instructions for the training of staff in safety and accident procedures.

In response to suggestions for epidemiological monitoring, the guidelines now require the principal investigator to report certain categories of accidents, in writing, to appropriate officials. NIH is investigating procedures for long-term surveillance of workers engaged in recombinant DNA research.

2. A number of comments on the role and responsibilities of the institutional biohazards committee were received. Comments were directed to the structure of the committee, the scope of its responsibility, and the methods for operation. Comments on structure included suggestions that the committee have a broadly based representation, especially in terms of health and safety expertise. Some others suggested NIH require certain classes of representation. In response to those suggestions, the guidelines now recommend membership from a diversity of disciplines relevant to recombinant DNA molecule technology, biological safety, and engineering.

For broader representation beyond the immediate scientific expertise, the guidelines now recommend that local committees should possess, or have available, the competence necessary to determine the acceptability of their findings in terms of applicable laws, regulations, standards of practice, community attitudes, and health and environmental considerations. The names of and relevant background information on the committee members will be reported to NIH.

In response to suggestions that decisions of the committee be made publicly available, the guidelines now recommend that minutes of the meetings should be kept and made available for public inspection.

Commentators generally approved of the responsibility given to the institutional biohazards committee to serve as a source of advice and reference to the investigator on scientific and safety questions. It was further suggested that the committee's responsibility be broadened in the development, monitoring, and evaluation of safety standards and procedures. In response to these suggestions, the guidelines now indicate that the institutional biohazards committee has the responsibility to certify, and recertify annually, to NIH that the facilities, procedures, practices, training, and expertise of involved personnel have been reviewed and approved. The Recombinant Advisory Committee suggested that examination might be unnecessary for P1 facilities, but we believe that all facilities should be reviewed to emphasize the importance of safety programs.

Some commentators suggested that the guidelines should stipulate that the local committees be required to determine the containment conditions to be imposed for a given project (which the draft guidelines specifically noted was not their responsibility). The Recombinant Advisory Committee took exception to this suggestion. They urged NIH not to include these conditions as local requirements, arguing among other things that review by the NIH study sections would provide the necessary scrutiny at the national level and assure uniformity of standards in application of the guidelines. I do not believe that NIH should require the local institution to have its biohazards committee assess what containment conditions are required for a given project. On the other hand, the guidelines should not prohibit the local institution from having its biohazards committee perform this function. Accordingly, I have deleted the prohibition that appeared in the proposed guidelines.

Another suggestion was that the local committee ensure that research is carried out in accordance with standards and procedures under the Occupational Safety and Health Act (OSHA). This is an area of importance to the local institutions under Federal and State law, but need not be included as a requirement in the guidelines. NIH will maintain liaison with the Occupational Safety and Health Administration (Department of Labor) to ensure maximum Federal cooperation in this venture.

I would also encourage all institutions, as suggested by several commentators, to review their insurance compensation programs to determine whether their laboratory personnel, in the research area, are covered for injuries.

3. The commentators approved of having the NIH study sections responsible for making an independent evaluation of the classification of the proposed research under the guidelines, along with the customary judgment of the scientific merit of each grant application. This additional element of review will ensure careful attention to potential hazards in the research activity. The study sections will also scrutinize the proposed safeguards. Biological safety expertise shall be available to the study section for consultation and guidance in this regard.

4. Several commentators made suggestions

concerning the structure, function, and scope of responsibility of the NIH Recombinant DNA Molecule Program Advisory Committee.

Comments on possible structural mechanisms for decision making included suggestions that there be a scientific and technical committee and a general advisory public policy committee. It was also suggested that the scientific committee include scientists who are not actively engaged in recombinant research, and that the public policy committee have a broad scientific and public representation.

I have carefully reviewed these comments and suggestions. In response, the following structure has been devised. The Recombinant Advisory Committee shall serve as the scientific and technical committee. Its membership shall continue to include scientists who represent disciplines actively engaged in recombinant DNA research. In my view, it is most important that this committee have the necessary expertise to assure that the guidelines are of the highest scientific quality. The committee has provided this expertise in the past, and it must continue to do so. The committee shall also include members from other scientific disciplines.

It should be noted that the present committee recommended on its own initiative that a nonscientist be appointed. Emmette S. Redford, Ph.D., LL.D., Ashbel Smith Professor of Government and Public Affairs at the Lyndon B. Johnson School of Public Affairs, University of Texas at Austin, serves in that capacity. An ethicist has also been nominated for appointment.

The Advisory Committee to the Director, NIH, shall serve to provide the broader public policy perspectives. This committee, at its meeting on February 9–10, 1976, reviewed the proposed guidelines with the participation of public witnesses, and shall continue to provide such review for future activities of the Recombinant Advisory Committee.

In response to suggestions, the responsibilities of the Recombinant Advisory Committee have been expanded. In addition to reviewing the guidelines for possible modification as scientific evidence warrants, the committee will certify EK2 and EK3 systems. In response to requests by the investigator, local committee, or study section, the committee will also provide evaluation and review in order to advise on levels of required containment, on lowering of requirements when cloned recombinants are to be used, and on questions concerning potential biohazard and adequacy of containment provisions.

Commentators also asked that the committee review ongoing research initiated prior to the implementation of the guidelines. Now that the guidelines are being released, NIH-funded investigators in this field will be asked to give assurance, within a given period, that they will comply. Any investigators who constructed clones under the Asilomar guidelines will be asked to petition NIH for special consideration of their case, if the new guidelines require higher containment than did the Asilomar guidelines. Here the advice of the Recombinant Advisory Committee will be sought.

There were also suggestions that the committee certify chemical purification of recombinant DNA, but as I indicated earlier, these procedures are too well known to require NIH monitoring.

5. In light of comments received, NIH will provide review, through appropriate NIH offices, of data from institutional biohazards committees (including accident reports) and will ensure dissemination of these findings as appropriate. Dr. William Gartland will head the newly created NIH Office of Recombinant DNA Activities for these purposes. In addition, NIH will provide for rapid dissemination of information through its Nucleic Acid Recombinant Scientific Memoranda (NARSM), distributed by the National Institute for Allergy and Infectious Diseases. NIH will also provide an appropriate mechanism for approving and certifying clones before containment conditions can be lowered.

With these extended modifications, the section of the guidelines dealing with roles and responsibilities now sets forth a more fully developed review structure involving the principal investigator, local biohazards committees, and the Recombinant Advisory Committee, as well as peer review committees, and the Recombinant Advisory Committee, as well as peer review committees. Guidelines now provide extensive opportunity for advice, from the local to the national level. Several levels of review and scrutiny are provided, ensuring the highest standards for scientific merit and conditions for safety.

The Recombinant Advisory Committee in conjunction with the Director's Advisory Committee shall continue to serve as an ongoing forum for examining progress in the technology and safety of recombinant DNA research. Their responsibility, and that of the NIH Director, is to ensure that the guidelines, through modification when called for, reflect the soundest scientific and safety evidence as it accrues in this area. Their task, in a sense, is just beginning.

DONALD S. FREDRICKSON,
Director, National Institute of Health

Guidelines for Research Involving Recombinant DNA Molecules, June 1976

The NIH guidelines (reprinted here without the appendices) were issued in June of 1976, and are subject to change on recommendation of the Recombinant DNA Molecule Advisory Committee to the NIH Director. Proposed revisions were published in the *Federal Register,* September 27, 1977. Reactions were received and a new proposal, together with a decision document by the Director, was published in the *Federal Register* on July 28, 1978 (Vol. 43, No. 146). A new cycle of response, including a public hearing before senior officers of the Department of Health, Education, and Welfare, is expected to delay formal promulgation of revised guidelines until late 1978 or early 1979.

I. INTRODUCTION

The purpose of these guidelines is to recommend safeguards for research on recombinant DNA molecules to the National Institutes of Health and to other institutions that support such research. In this context we define recombinant DNAs as molecules that consist of different segments of DNA which have been joined together in cell-free systems, and which have the capacity to infect and replicate in some host cell, either autonomously or as an integrated part of the host's genome.

This is the first attempt to provide a detailed set of guidelines for use by study sections as well as practicing scientists for evaluating research on recombinant DNA molecules. We cannot hope to anticipate all possible lines of imaginative research that are possible with this powerful new methodology. Nevertheless, a considerable volume of written and verbal contributions from scientists in a variety of disciplines has been received. In many instances the views presented to us were con-

Federal Register, Vol. 41, No. 131, July 1976, pp. 27911–27922.

tradictory. At present, the hazards may be guessed at, speculated about, or voted upon, but they cannot be known absolutely in the absence of firm experimental data — and, unfortunately, the needed data were, more often than not, unavailable. Our problem then has been to construct guidelines that allow the promise of the methodology to be realized while advocating the considerable caution that is demanded by what we and others view as potential hazards.

In designing these guidelines we have adopted the following principles, which are consistent with the general conclusions that were formulated at the International Conference on Recombinant DNA Molecules held at Asilomar Conference Center, Pacific Grove, California, in February 1975 (3): (i) There are certain experiments for which the assessed potential hazard is so serious that they are not to be attempted at the present time. (ii) The remainder can be undertaken at the present time provided that the experiment is justifiable on the basis that new knowledge or benefits to humankind will accrue that cannot readily be obtained by use of conventional methodology and that appropriate safeguards are incorporated into the design and execution of the experiment. In addition to an insistence on the practice of good microbiological techniques, these safeguards consist of providing both physical and biological barriers to the dissemination of the potentially hazardous agents. (iii) The level of containment provided by these barriers is to match the estimated potential hazard for each of the different classes of recombinants. For projects in a given class, this level is to be highest at initiation and modified subsequently only if there is a substantiated change in the assessed risk or in the applied methodology. (iv) The guidelines will be subjected to periodic review (at least annually) and modified to reflect improvements in our knowledge of the potential biohazards and of the available safeguards.

In constructing these guidelines it has been necessary to define boundary conditions for the different levels of physical and biological containment and for the classes of experiments to which they apply. We recognize that these definitions do not take into account existing and anticipated special procedures and information that will allow particular experiments to be carried out under different conditions than indicated here without sacrifice of safety. Indeed, we urge that individual investigators devise simple and more effective containment procedures and that study sections give consideration to such procedures which may allow change in the containment levels recommended here.

It is recommended that all publications dealing with recombinant DNA work include a description of the physical and biological containment procedures practiced, to aid and forewarn others who might consider repeating the work.

II. CONTAINMENT

Effective biological safety programs have been operative in a variety of laboratories for many years. Considerable information therefore already exists for the design of physical containment facilities and the selection of laboratory procedures applicable to organisms carrying recombinant DNAs (4–17). The existing programs rely upon mechanisms that, for convenience, can be divided into two categories: (i) a set of standard practices that are generally used in microbiological laboratories, and (ii) special procedures, equipment, and laboratory installations that provide physical barriers which are applied in varying degrees according to the estimated biohazard.

Experiments on recombinant DNAs by their very nature lend themselves to a third containment mechanism — namely, the application of highly specific biological barriers. In fact, natural barriers do exist which either limit the infectivity of a vector or vehicle (plasmid, bacteriophage or virus) to specific hosts, or its dissemination and survival in the environment. The vectors that provide the means for replication of the recombinant DNAs and/or the host cells in which they replicate can be genetically designed to decrease by many orders of magnitude the probability of dissemination of recombinant DNAs outside the laboratory.

As these three means of containment are complementary, different levels of containment appropriate for experiments with different recombinants can be established by applying different combinations of the physical and biological barriers to a constant use of the standard practices. We consider these categories of containment separately here in order that such combinations can be conveniently expressed in the guidelines for research on the different kinds of recombinant DNA (Section III).

A. STANDARD PRACTICES AND TRAINING

The first principle of containment is a strict adherence to good microbiological practices (4–13). Consequently, all personnel directly or indirectly involved in experiments on recombinant DNAs must receive adequate instruction. This should include at least training in aseptic techniques and instruction in the biology of the organisms used in the experiments so that the potential biohazards can be understood and appreciated.

Any research group working with agents with a known or potential biohazard should have an emergency plan which describes the procedures to be followed if an accident contaminates personnel or environment. The principal investigator must ensure that everyone in the laboratory is familiar with both the potential hazards of the work and the emergency plan. If a research group is working with a known pathogen for which an effective vaccine is available, all workers should be immunized. Serological monitoring, where appropriate, should be provided.

B. PHYSICAL CONTAINMENT LEVELS

A variety of combinations (levels) of special practices, equipment, and laboratory installations that provide additional physical barriers can be formed. For example, 31 combinations are listed in "Laboratory Safety at the Center for Disease Control" (4); four levels are associated with the Classification of Etiologic Agents on the Basis of Hazard" (5), four levels were recommended in the "Summary Statement of the Asilomar Conference on Recombinant DNA Molecules" (3); and the National Cancer Institute uses three levels for research on oncogenic viruses (6). We emphasize that these are an aid to, and not a substitute for, good technique. Personnel must be competent in the effective use of all equipment needed for the required containment level as described below. We define only four levels of physical containment here, both because the accuracy with which one can presently assess the biohazards that may result from recombinant DNAs does not warrant a more detailed classification, and because additional flexibility can be obtained by combination of the physical with the biological barriers. Though different in detail, these four levels (P1<P2<P3<P4) approximate those given for human etiologic agents by the Center for Disease Control (i.e., classes 1 through 4; ref. 5), in the Asilomar summary statement (i.e., minimal, low, moderate, and high; ref. 3), and by the National Cancer Institute for oncogenic viruses (i.e., low, moderate, and high; ref. 6), as is indicated by the P-number or adjective in the following headings. It should be emphasized that the descriptions and assignments of physical containment detailed below are based on existing approaches to containment of hazardous organisms.

We anticipate, and indeed already know of, procedures (14) which enhance physical containment capability in novel ways. For example, miniaturization of screening, handling, and analytical procedures provides substantial containment of a given host-vector system. Thus, such procedures should reduce the need for the standard types of physical containment, and such innovations will be considered by the Recombinant DNA Molecule Program Advisory Committee.

The special practices, equipment and facility installations indicated for each level of physical containment are required for the safety of laboratory workers, other persons, and for the protection of the environment. Optional items have been excluded; only those items deemed absolutely necessary for safety are presented. Thus, the listed requirements present basic safety criteria for each level of physical containment. Other microbiological practices and laboratory techniques which promote safety are to be encouraged. Additional information giving further guidance on physical containment is provided in a supplement to the guidelines (Appendix D).

P1 Level (Minimal)   A laboratory suitable for experiments involving recombinant DNA molecules requiring physical containment at the P1 level is a laboratory that possesses no special engineering design features. It is a laboratory commonly used for microorganisms of no or minimal biohazard under ordinary conditions of handling. Work in this laboratory is generally conducted on open bench tops. Special containment equipment is neither required nor generally

available in this laboratory. The laboratory is not separated from the general traffic patterns of the building. Public access is permitted.

The control of biohazards at the P1 level is provided by standard microbiological practices of which the following are examples: (i) Laboratory doors should be kept closed while experiments are in progress. (ii) Work surfaces should be decontaminated daily and following spills of recombinant DNA materials. (iii) Liquid wastes containing recombinant DNA materials should be decontaminated before disposal. (iv) Solid wastes contaminated with recombinant DNA materials should be decontaminated or packaged in a durable leak-proof container before removal from the laboratory. (v) Although pipetting by mouth is permitted, it is preferable that mechanical pipetting devices be used. When pipetting by mouth, cotton-plugged pipettes shall be employed. (vi) Eating, drinking, smoking, and storage of food in the working area should be discouraged. (vii) Facilities to wash hands should be available. (viii) An insect and rodent control program should be provided. (ix) The use of laboratory gowns, coats, or uniforms is discretionary with the laboratory supervisor.

P2 Level (Low)   A laboratory suitable for experiments involving recombinant DNA molecules requiring physical containment at the P2 level is similar in construction and design to the P1 laboratory. The P2 laboratory must have access to an autoclave within the building; it may have a Biological Safety Cabinet.* Work which does not produce a considerable aerosol is conducted on the open bench. Although this laboratory is not separated from the general traffic patterns of the building, access to the laboratory is limited when experiments requiring P2 level physical containment are being conducted. Experiments of lesser biohazard potential can be carried out concurrently in carefully demarcated areas of the same laboratory.

The P2 laboratory is commonly used for experiments involving microorganisms of low biohazard such as those which have been classified by the Center for Disease Control as Class 2 agents (5).

The following practices shall apply to all experiments requiring P2 level physical containment: (i) Laboratory doors shall be kept closed while experiments are in progress. (ii) Only persons who have been advised of the potential biohazard shall enter the laboratory. (iii) Children under 12 years of age shall not enter the laboratory. (iv) Work surfaces shall be decontaminated daily and immediately following spills of recombinant DNA materials. (v) Liquid wastes of recombinant DNA materials shall be decontaminated before disposal. (vi) Solid wastes contaminated with recombinant DNA materials shall be decontaminated or packaged in a durable leak-proof container before removal from the laboratory. Packaged materials shall be disposed of by incineration or sterilized before disposal by other methods. Contaminated materials that are to be processed and reused (i.e., glassware) shall be decontaminated before removal from the laboratory. (vii) Pipetting by mouth is prohibited; mechanical pipetting devices shall be used. (viii) Eating, drinking, smoking, and storage of food are not permitted in the working area. (ix) Facilities to wash hands shall be available within the laboratory. Persons handling recombinant DNA materials should be encouraged to wash their hands frequently and when they leave the laboratory. (x) An insect and rodent control program shall be provided. (xi) The use of laboratory gowns, coats, or uniforms is required. Such clothing shall not be worn to the lunch room or outside the building. (xii) Animals not related to the experiment shall not be permitted in the laboratory. (xiii) Biological Safety Cabinets[1] and/or other physical containment equipment shall be used to minimize the hazard of aerosolization of recombinant DNA materials from operations or devices that produce a considerable aerosol (e.g., blender, lyophilizer, sonicator, shaking machine, etc.). (xiv) Use of the hypodermic needle and syringe shall be avoided when alternate methods are available.

P3 Level (Moderate) A laboratory suitable for experiments involving recombinant DNA molecules requiring physical containment at the P3 level has special engineering design features and physical containment equipment. The laboratory is separated from areas which are open to the general public. Separation is generally achieved by controlled access corridors, air locks, locker rooms or other double-doored facilities which are not available for use by the general public. Access to the laboratory is controlled. Biological Safety Cabinets[1] are available within the controlled laboratory area. An autoclave shall be available within the building and preferably within the controlled laboratory area. The surfaces of walls, floors, bench tops, and ceilings are easily cleanable to facilitate housekeeping and space decontamination.

Directional air flow is provided within the controlled laboratory area. The ventilation system is balanced to provide for an inflow of supply air from the access corridor into the laboratory. The general exhaust air from the laboratory is discharged outdoors and so dispersed to the atmosphere as to prevent reentry into the building. No recirculation of the exhaust air shall be permitted without appropriate treatment.

*Footnotes at end of article.

No work in open vessels involving hosts or vectors containing recombinant DNA molecules requiring P3 physical containment is conducted on the open bench. All such procedures are confined to Biological Safety Cabinets.[1]

The following practices shall apply to all experiments requiring P3 level physical containment: (i) The universal biohazard sign is required on all laboratory access doors. Only persons whose entry into the laboratory is required on the basis of program or support needs shall be authorized to enter. Such persons shall be advised of the potential biohazards before entry and they shall comply with posted entry and exit procedures. Children under 12 years of age shall not enter the laboratory. (ii) Laboratory doors shall be kept closed while experiments are in progress. (iii) Biological Safety Cabinets[1] and other physical containment equipment shall be used for all procedures that produce aerosols of recombinant DNA materials (e.g., pipetting, plating, flaming, transfer operations, grinding, blending, drying, sonicating, shaking, etc.). (iv) The work surfaces of Biological Safety Cabinets[1] and other equipment shall be decontaminated following the completion of the experimental activity contained within them. (v) Liquid wastes containing recombinant DNA materials shall be decontaminated before disposal. [vi] Solid wastes contained with recombinant DNA materials shall be decontaminated or packaged in a durable leak-proof container before removal from the laboratory. Packaged material shall be sterilized before disposal. Contaminated materials that are to be processed and reused (i.e., glassware) shall be sterilized in the controlled laboratory area or placed in a durable leak-proof container before removal from the controlled laboratory area. This container shall be sterilized before the materials are processed. (vii) Pipetting by mouth is prohibited; mechanical pipetting devices shall be used. [viii] Eating, drinking, smoking, and storage of food are not permitted in the laboratory. (ix) Facilities to wash hands shall be available within the laboratory. Persons shall wash hands after experiments involving recombinant DNA materials and before leaving the laboratory. (x) An insect and rodent control program shall be provided. (xi) Laboratory clothing that protects street clothing (i.e., long sleeve solid-front or wrap-around gowns, no-button or slipover jackets, etc.) shall be worn in the laboratory. FRONT-BUTTON LABORATORY COATS ARE UNSUITABLE. Gloves shall be worn when handling recombinant DNA materials. Provision for laboratory shoes is recommended. Laboratory clothing shall not be worn outside the laboratory and shall be decontaminated before it is sent to the laundry. (xii) Raincoats, overcoats, topcoats, coats, hats, caps, and such street outerwear shall not be kept in the laboratory.

(xiii) Animals and plants not related to the experiment shall not be permitted in the laboratory. (xiv) Vacuum lines shall be protected by filters and liquid traps. (xv) Use of the hypodermic needle and syringe shall be avoided when alternate methods are available. (xvi) If experiments of lesser biohazard potential are to be conducted in the same laboratory concurrently with experiments requiring P3 level physical containment they shall be conducted only in accordance with all P3 level requirements. (xvii) Experiments requiring P3 level physical containment can be conducted in laboratories where the directional air flow and general exhaust air conditions described above cannot be achieved, provided that this work is conducted in accordance with all other requirements listed and is contained in a Biological Safety Cabinet[1] with attached glove ports and gloves. All materials before removal from the Biological Safety Cabinet[1] shall be sterilized or transferred to a non-breakable, sealed container, which is then removed from the cabinet through a chemical decontamination tank, autoclave, ultraviolet air lock, or after the entire cabinet has been decontaminated.

P4 Level (High)   Experiments involving recombinant DNA molecules requiring physical containment at the P4 level shall be confined to work areas in a facility of the type designed to contain microorganisms that are extremely hazardous to man or may cause serious epidemic disease. The facility is either a separate building or it is a controlled area, within a building, which is completely isolated from all other areas of the building. Access to the facility is under strict control. A specific facility operations manual is available. Class III Biological Safety Cabinets[1] are available within work areas of the facility.

A P4 facility has engineering features which are designed to prevent the escape of microorganisms to the environment (14, 15, 16, 17). These features include: (i) Monolithic walls, floods, and ceilings in which all penetrations such as for air ducts, electrical conduits, and utility pipes are sealed to assure the physical isolation of the work area and to facilitate housekeeping and space decontamination; (ii) air locks through which supplies and materials can be brought safely into the facility; (iii) contiguous clothing change and shower rooms through which personnel enter into and exit from the facility; (iv) double-door autoclaves to sterilize and safely remove wastes and other materials from the facility; (v) a biowaste treatment system to sterilize liquid effluents if facility drains are installed; (vi) a separate ventilation system which maintains negative air pressures and directional air flow within the facility; and (vii) a treatment system to decontaminate exhaust air before it is dispersed to the atmosphere. A central vacuum utility system is not encouraged; if one is

installed, each branch line leading to a laboratory shall be protected by a high efficiency particulate air filter.

The following practices shall apply to all experiments requiring P4 level physical containment: (i) The universal biohazard sign is required on all facility access doors and all interior doors to individual laboratory rooms where experiments are conducted. Only persons whose entry into the facility or individual laboratory rooms is required on the basis of program or support needs shall be authorized to enter. Such persons shall be advised of the potential biohazards and instructed as to the appropriate safeguards to ensure their safety before entry. Such persons shall comply with the instructions and all other posted entry and exit procedures. Under no condition shall children under 15 years of age be allowed entry. (ii) Personnel shall enter into and exit from the facility only through the clothing change and shower rooms. Personnel shall shower at each exit from the facility. The air locks shall not be used for personnel entry or exit except for emergencies. (iii) Street clothing shall be removed in the outer facility side of the clothing change area and kept there. Complete laboratory clothing including undergarments, pants and shirts or jumpsuits, shoes, head cover, and gloves shall be provided and used by all persons who enter into the facility. Upon exit, this clothing shall be stored in lockers provided for this purpose or discarded into collection hampers before personnel enter into the shower area. (iv) Supplies and materials to be taken into the facility shall be placed in an entry air lock. After the outer door (opening to the corridor outside of facility) has been secured, personnel occupying the facility shall retrieve the supplies and materials by opening the interior air lock door. This door shall be secured after supplies and materials are brought into the facility. (v) Doors to laboratory rooms within the facility shall be kept closed while experiments are in progress. (vi) Experimental procedures requiring P4 level physical containment shall be confined to Class III Biological Safety Cabinets.[1] All materials, before removal from these cabinets, shall be sterilized or transferred to a non-breakable sealed container, which is then removed from the system through a chemical decontaminated tank, autoclave, or after the entire system has been decontaminated.

(vii) No materials shall be removed from the facility unless they have been sterilized or decontaminated in a manner to prevent the release of agents requiring P4 physical containment. All wastes and other materials and equipment not damaged by high temperature or steam shall be sterilized in the double-door autoclave. Biological materials to be removed from the facility shall be transferred to a non-breakable sealed container which is then removed from the facility through a chemical decontamination tank or a chamber designed for gas sterilization. Other materials which may be damaged by temperature or steam shall be sterilized by gaseous or vapor methods in an air lock or chamber designed for this purpose. (viii) Eating, drinking, smoking, and storage of food are not permitted in the facility. Foot-operated water fountains located in the facility corridors are permitted. Separate potable water piping shall be provided for these water fountains. (ix) Facilities to wash hands shall be available within the facility. Persons shall wash hands after experiments. (x) An insect and rodent control program shall be provided. (xi) Animals and plants not related to the experiment shall not be permitted in the facility. (xii) If a central vacuum system is provided, each vacuum outlet shall be protected by a filter and liquid trap in addition to the branch line HEPA filter mentioned above. (xiii) Use of the hypodermic needle and syringe shall be avoided when alternate methods are available. (xiv) If experiments of lesser biohazard potential are to be conducted in the facility concurrently with experiments requiring P4 level containment, they shall be confined in Class I or Class II Biological Safety Cabinets[1] or isolated by other physical containment equipment. Work surfaces of Biological Safety Cabinets[1] and other equipment shall be decontaminated following the completion of the experimental activity contained within them. Mechanical pipetting devices shall be used. All other practices listed above with the exception of (vi) shall apply.

C. SHIPMENT

To protect product, personnel, and the environment, all recombinant DNA material will be shipped in containers that meet the requirements issued by the U.S. Public Health Service (Section 72.25 of Part 72, Title 42, Code of Federal Regulations), Department of Transportation (Section 173.387(b) of Part 173, Title 49, Code of Federal Regulations) and the Civil Aeronautics Board (C.A.B. No. 82, Official Air Transport Restricted Articles Tariff No. 6—D) for shipment of etiologic agents. Labelling requirements specified in these Federal regulations and tariffs will apply to all viable recombinant DNA materials in which any portion of the material is derived from an etiologic agent listed in paragraph (c) of 42 CFR 72.25. Additional information on packing and shipping is given in a supplement to the guidelines (Appendix D, part X).

D. BIOLOGICAL CONTAINMENT LEVELS

Biological barriers are specific to each host-vector system. Hence the criteria for this mechanism of containment cannot be generalized to the same extent as for physical containment. This is particularly true at the present time when our experience with existing host-vector systems and our predictive knowledge about projected systems are sparse. The classification of experiments with recombinant DNAs that is necessary for the construction of the experimental guidelines (Section III) can be accomplished with least confusion if we use the host-vector system as the primary element and the source of the inserted DNA as the secondary element in the classification. It is therefore convenient to specify the nature of the biological containment under host-vector headings such as those given below for *Escherichia coli* K–12.

III. EXPERIMENTAL GUIDELINES

A general rule that, though obvious, deserves statement is that the level of containment required for any experiment on DNA recombinants shall never be less than that required for the most hazardous component used to construct and clone the recombinant DNA (i.e., vector, host, and inserted DNA). In most cases the level of containment will be greater, particularly when the recombinant DNA is formed from species that ordinarily do not exchange genetic information. Handling the purified DNA will generally require less stringent precautions than will propagating the DNA. However, the DNA itself should be handled at least as carefully as one would handle the most dangerous of the DNAs used to make it.

The above rule by itself effectively precludes certain experiments — namely, those in which one of the components is in Class 5 of the "Classification of Etiologic Agents on the Basis of Hazard" (5), as these are excluded from the United States by law and USDA administrative policy. There are additional experiments which may engender such serious biohazards that they are not to be performed at this time. These are considered prior to presentation of the containment guidelines for permissible experiments.

A. EXPERIMENTS THAT ARE NOT TO BE PERFORMED

We recognize that it can be argued that certain of the recombinants placed in this category could be adequately contained at this time. Nonetheless, our estimates of the possible dangers that may ensue if that containment fails are of such a magnitude that we consider it the wisest policy to at least defer experiments on these recombinant DNAs until there is more information to accurately assess that danger and to allow the construction of more effective biological barriers. In this respect, these guidelines are more stringent than those initially recommended (1).

The following experiments are not to be initiated at the present time: (1) Cloning of recombinant DNAs derived from the pathogenic organisms in Classes 3, 4, and 5 of "Classification of Etiologic Agents on the Basis of Hazard" (5), or oncogenic viruses classified by NCI as moderate risk (6), or cells known to be infected with such agents, regardless of the host-vector system used. (ii) Deliberate formation of recombinant DNAs containing genes for the biosynthesis of potent toxins (e.g., botulinum or diphtheria toxins; venoms from insects, snakes, etc.). (iii) Deliberate creation from plant pathogens of recombinant DNAs that are likely to increase virulence and host range. (iv) Deliberate release into the environment of any organism containing a recombinant DNA molecule. (v) Transfer of a drug resistance trait to microorganisms that are not known to acquire it naturally if such acquisition could compromise the use of a drug to control disease agents in human or veterinary medicine or agriculture.

In addition, at this time large-scale experiments (e.g., more than 10 liters of culture) with recombinant DNAs known to make harmful product are not to be carried out. We differentiate between small- and large-scale experiments with such DNAs because the probability of escape from containment barriers normally increases with increasing scale. However, specific experiments in this category that are of direct societal benefit may be excepted from this rule if special biological containment precautions and equipment designed for large-scale operations are used, and provided that these experiments are expressly approved by the Recombinant DNA Molecule Program Advisory Committee of NIH.

B. CONTAINMENT GUIDELINES FOR PERMISSIBLE EXPERIMENTS

It is anticipated that most recombinant DNA experiments initiated before these guidelines are next reviewed (i.e., within the year) will employ $E.$ $coli$ K– 12 host-vector systems. These are also the systems for which we have the most experience and knowledge regarding the effectiveness of the containment provided by existing hosts and vectors necessary for the construction of more effective biological barriers.

For these reasons, $E.$ $coli$ K– 12 appears to be the system of choice at this time, although we have carefully considered arguments that many of the potential dangers are compounded by using an organism as intimately connected with a man as is $E.$ $coli.$ Thus, while proceeding cautiously with $E.$ $coli,$ serious efforts should be made toward developing alternate host-vector systems; this subject is discussed in considerable detail in Appendix A.

We therefore consider DNA recombinants in $E.$ $coli$ K– 12 before proceeding to other host-vector systems.

1. Biological containment criteria using E. coli *K– 12 host-vectors —EK1 host-vectors* These are host-vector systems that can be estimated to already provide a moderate level of containment, and include most of the presently available systems. The host is always $E.$ $coli$ K– 12, and the vectors include nonconjugative plasmids [e.g., pSC101, ColE1 or derivatives thereof (19– 26)] and variants of bacteriophage λ (27– 29).

The $E.$ $coli$ K– 12 nonconjugative plasmid system is taken as an example to illustrate the approximate level of containment referred to here. The available data from experiments involving the feeding of bacteria to humans and calves (30– 32) indicate that $E.$ $coli$ K– 12 did not usually colonize the *normal* bowel, and exhibited little, if any, multiplication while passing through the alimentary tract even after feeding high doses (i.e., 10^9 to 10^{10} bacteria per human or calf). However, general extrapolation of these results may not be warranted because the implantation of bacteria into the intestinal tract depends on a number of parameters, such as the nature of the intestinal flora present in a given individual and the physiological state of the inoculum. Moreover, since viable $E.$ $coli$ K– 12 can be found in the feces after humans are fed 10^7 bacteria in broth (30) or 3×10^4 bacteria protected by suspension in milk (31), transductional and conjugational transfer of the plasmid vectors from $E.$ $coli$ K– 12 to resident bacteria in the fecal matter before and after excretion must also be considered.

The nonconjugative plasmid vectors cannot promote their own transfers, but require the presence of a conjugative plasmid for mobilization and transfer to other bacteria. When present in the same cell with derepressed conjugative plasmids such as F or R1*drd19,* the nonconjugative ColE1, ColE1-*trp* and pSC101 plasmids are transferred to suitable recipient strains under ideal laboratory conditions at frequencies of about 0.5, 10^{-4} to 10^{-5}, and 10^{-6} per donor cell, respectively. These frequencies are reduced by another factor of 10^2 to 10^4 if the conjugative plasmid employed is repressed with respect to expression of donor fertility.

The experimental transfer system which most closely resembles nonconjugative plasmid transfer in nature is a triparental mating. In such matings, the bacterial cell possessing the nonconjugative plasmid must first acquire a conjugative plasmid from another cell before it can transfer the nonconjugative plasmid to a secondary recipient. With ColE1, the frequencies of transfer are 10^{-2} and 10^{-4} to 10^{-5} when using conjugative plasmid donors possessing derepressed and repressed plasmids, respectively. Mobilization of ColE1-*trp* and pSC101 under similar laboratory conditions is so low as to be usually undetectable (33). Since most conjugative plasmids in nature are repressed for expression of donor fertility, the frequency at which nonconjugative plasmids are mobilized and transferred by this sequence of events *in vivo* is difficult to estimate. However, in calves fed on an antibiotic-supplemented diet, it has been estimated that such triparental nonconjugative R plasmid transfer occurs at frequencies of no more than 10^{-10} to 10^{-12} per 24 hours per calf (32). In terms of considering other means for plasmid transmission in nature, it should be noted that transduction does operate *in vivo* for *Staphylococcus aureus* (34) and probably for *E. coli* as well. However, no data are available to indicate the frequencies of plasmid transfer *in vivo* by either transduction or transformation.

These observations indicate the low probabilities for possible dissemination of such plasmid vectors by accidental ingestion, which would probably involve only a few hundred or thousand bacteria provided that at least the standard practices (Section II–A above) are followed, particularly the avoidance of mouth pipetting. The possibility of colonization and hence of transfer are increased, however, if the normal flora in the bowel is disrupted by, for example, antibiotic therapy (35). For this reason, persons receiving such therapy must not work with DNA recombinants formed with any *E. coli* K–12 host-vector system during the therapy period and for seven days thereafter; similarly, persons who have achlorhydria or who have had surgical removal of part of the stomach or bowel should avoid such work, as should those who require large doses of antacids.

The observations on the fate of *E. coli* K–12 in the human alimentary tract are also relevant to the containment of recombinant DNA formed with bacteriophage λ variants. Bacteriophage can escape from the laboratory either as mature infectious phage particles or in bacterial host cells in which the phage genome is carried as a plasmid or prophage. The fate of *E. coli* K–12 host cells carrying the phage genome as a plasmid or prophage is similar to that for plasmid-containing host cells as discussed above. The survival of the λ phage genome when released as infectious particles depends on their stability in nature, their infectivity and on the probability of subsequent encounters with naturally occurring λ-sensitive *E. coli* strains. Although the probability of survival of λ and its infection of resident intestinal *E. coli* in animals and humans has not been measured, it is estimated to be small given the high sensitivity of λ to the low pH of the stomach, the insusceptibility to λ infection of smooth *E. coli* cells (the type that normally resides in the gut), the infrequency of naturally occurring λ-sensitive *E. coli* (36) and the failure to detect infective λ particles in human feces after ingestion of up to 10^{11} λ particles (37). Moreover, λ particles are very sensitive to desiccation.

Establishment of λ as a stable lysogen is a frequent event (10^{0} to 10^{-1}) for the *att*$^{+}$ *int*$^{+}$ *cI*$^{+}$ phage so that this mode of escape would be the preponderant laboratory hazard; however, most EK1 λ vectors currently in use lack the *att* and *int* functions (27–29) thus reducing the probability of lysogenization to about 10^{-5} to 10^{-6} (38–40). The frequency for the conversion of λ to a plasmid state for persistence and replication is also only about 10^{-6} (41). Moreover, the routine treatment of phage lysates with chloroform (42) should eliminate all surviving bacteria including lysogens and λ plasmid carriers. Lysogenization could also occur when an infectious λ containing cloned DNA infects a λ-sensitive cell in nature, and recombines with a resident lambdoid prophage. Although λ-sensitive *E. coli* strains seem to be rare, a significant fraction do carry lambdoid prophages (43–44) and thus this route of escape should be considered.

While not exact, the estimates for containment afforded by using these host-vectors are at least as accurate as those for physical containment, and are sufficient to indicate that currently employed plasmid and λ vector systems provide a moderate level of biological containment. Other nonconjugative plasmids and bacteriophages that, in association with *E. coli* K–12 can be estimated to provide the same approximate level of moderate containment are included in the EK1 class.

EK2 host-vectors. These are host-vector systems that have been genetically constructed and shown to provide a high level of biological containment as demonstrated by data from suitable tests performed in the laboratory. The genetic modifications of the *E. coli* K–12 host and/or the plasmid or phage vector should not permit survival of a genetic marker carried on the vector, preferably a marker within an inserted DNA fragment, in other than specially designed and carefully regulated laboratory environments at a frequency greater than 10^{-8}. This measure of biological containment has been selected because it is a measurable entity. Indeed, by testing the contributions of preexisting and newly introduced genetic properties of vectors and hosts, individually or in various combinations, it should be possible to estimate with considerable precision, that the specially designed host-vector system can provide a margin of biological containment in excess of that required. For the time being, no host-vector system will be considered to be a bona fide EK2 host-vector system until it is so certified by the NIH Recombinant DNA Molecule Program Advisory Committee.

For EK2 host-vector systems in which the vector is a plasmid, no more than one in 10^8 host cells should be able to perpetuate the vector and/or a cloned DNA fragment under non-permissive conditions designed to represent the natural environment either by survival of the original host or as a consequence of transmission of the vector and/or a cloned DNA fragment by transformation, transduction or conjugation to a host with properties common to those in the natural environment.

In terms of potential EK2 plasmid-host systems, the following types of genetic modifications should reduce survival of cloned DNA. *The examples given are for illustrative purposes and should not be construed to encompass all possibilities.* The presence of the non-conjugative plasmids ColE1-*trp* and pS101 in an *E. coli* K–12 strain possessing a mutation eliminating host-controlled restriction and modification (*hsdS*) results in about 10^2-fold reduction in mobilization to restriction-proficient recipients. The combination of *dapD8*, Δ*bioH-asd*, Δ*gal-chl*^{r} and *rfb* mutations in *E. coli* K–12 results in no detectable survivors in feces of rats following feeding by stomach tube of 10^{10} cells in milk and similarly leads to complete lysis of cells suspended in broth medium lacking diaminopimelic acid. *E. coli* K–12 strains with Δ*thyA* and *deoC* (*dra*) mutations undergo thymineless death in growth medium lacking thymine and give a 10^5-fold reduced survival during passage through the rat intestine compared to wild-type *thy*$^+$ *E. coli* K–12. (However, the Δ*thyA* mutation alone or in combination with a *deoB*(*drm*) mutation only reduces *in vivo* survival by a factor of 10^2.) Other host mutations, as yet untested, that might further reduce survival of the plasmid-host system or reduce plasmid transmission are: the combination *polA*(TS) *recA*(TS) Δ*thyA* which might interfere with ColE1 replication and lead to DNA degradation at body temperatures; Con– mutations that reduce the ability of conjugative plasmids to enter the plasmid-host complex and thus should reduce mobilization of the cloned DNA to other strains; and mutations that confer resistance to known transducing phages. Mutations can also be introduced into the plasmid to cause it to be dependent on a specific host, to make its replication thermosensitive and/or to endow it with a kill-er capability such that all cells (other than its host) into which it might be transferred with not survive.

In the construction of EK2 plasmid-host systems it is important to use the most stable mutations available, preferably deletions. Obviously, the presence of all mutations contributing to higher degrees of biological containment must be verified periodically by appropriate tests. In testing the level of biological containment afforded by a proposed EK2 plasmid-host system, it is important to design relevant tests to evaluate the survival of the vector and/or a cloned DNA fragment under conditions that are possible in nature and that are also most advantageous for its perpetuation. For example, one might conduct a triparental mating with a primary donor possessing a derepressed F-type or I-type conjugative plasmid, the safer host with Δ*bioH-asd*, *dapD8*, Δ*gal-chl*^{r}, *rfb*, Δ*thyA*, *deoC*, *trp* and *hsdS* mutations and a plasmid vector carrying an easily detectable marker such as for ampicillin resistance or an inserted gene such as *trp*$^+$, and a secondary recipient that is Su$^+$ *hsdS trp* (i.e., permissive for the recombinant plasmid). Such matings would be conducted in a medium lacking diaminopimelic acid and thymine and survival of the Ap^r or *trp*$^+$ marker in any of the three strains followed as a function of time. Survival of the vector and/or a cloned marker by transduction could also be evaluated by introducing a known generalized transducing phage into the system. Similar experiments should also be done using a secondary recipient that is restrictive for the plasmid vector as well as with primary donors possessing repressed conjugative plasmids with incompatibility group properties like those commonly found in enteric microorganisms. Since a common route of escape of plasmid-host systems in the laboratory might be by accidental ingestion, it is suggested that the same types of experiments be conducted in suitable animal-model systems. In addition to these tests on survival of the vector and/or a cloned DNA fragment, it would be useful to determine the survival of the host strain under nongrowth conditions such as in water and as a function of drying time after a culture has been spilled on a lab bench.

For EK2 host-vector systems in which the vector is a phage, no more than one in 10^8 phage particles should be able to perpetuate itself and/or a cloned DNA fragment under non-permissive conditions designed to represent the natural environment either (a) as a prophage or plasmid in the laboratory host used for phage propagation or (b) by surviving in natural environments and transferring itself and/or a cloned DNA fragment to a host (or its resident lamboid prophage) with properties common to those in the natural environment.

In terms of potential EK2 λ-host systems, the following types of genetic modification should reduce survival of cloned DNA. *The examples given are for illustrative purposes and should not be construed to encompass all possibilities.* The probability of establishing λ lysogeny in the normal laboratory host should be reduced by removal of the phage *att* site, the Int function, the repressor gene(s) and adding virulence-enhancing mutations. The frequency of plasmid formation, although normally already less than 10^{-6}, could be further reduced by defects in the p_R-Q region, including mutations such as *vir-s*, *cro*(TS), c^{17}, ri^{cee}, O(TS), P(TS), and *nin*. Moreover, chloroform treatment used routinely following cell lysis would reduce the number of surviving cells, including possible lysogens or plasmid carriers, by more than 10^8. The host may also be modified by deletion of the host λ*att* site and inclusion of one or more of the mutations described above for plasmid-host systems to further reduce the chance of formation and survival of any lysogen or plasmid carrier cell.

The survival of escaping phage and the chance of encountering a sensitive host in nature are very low, as discussed for EK1 systems. The infectivity of the phage particles could be further reduced by introducing mutations (e.g., suppressed ambers) which would make the phage particles extremely unstable except under special laboratory conditions (e.g., high concentrations of salts or putrescine). Another means would be to make the phage itself a two-component system, by eliminating the tail genes and reproducing the phage as heads packed with DNA; when necessary and under specially controlled conditions, these heads could be made infective by adding tail preparations. An additional safety factor in this regimen is the extreme instability of the heads, unless they are stored in 10mM putrescine, a condition easy to obtain in the laboratory but not in nature. The propagation of the escaping phage in nature could further be blocked by adding various conditional mutations which would permit growth only under special laboratory conditions or in a special permissive laboratory host with suppressor or *gro*-type (*mop*, *dnaB*, *rpoB*) mutations. An additional safety feature would be the use of an r^+m^+ (*hsdS*) laboratory host, which produces phage with unmodified DNA which should be restricted in r^+m^+ bacteria that are probably prevalent in nature. The likelihood of recombination between the λ vector and lambdoid prophages which are present in some *E. coli* strains might be reduced by elimination of the Red function and the presence of the recombination-reducing Gam function together with mutations contributing to the high lethality of the λ phage. However, these second-order precautions might not be relevant if the stability and infectivity of the escaping λ particles are reduced by special mutations or by propagating the highly unstable heads.

Despite multiple mutations in the phage vectors and laboratory hosts, the yield of phage particles under suitable laboratory conditions should be high (10^{10}–10^{11} particles/ml). This permits phage propagation in relatively small volumes and constitutes an additional safety feature.

The phenotypes and genetic stabilities of the mutations and chromosome alterations included in these λ-host systems indicate that containment well in excess of the required 10^{-8} or lower survival frequency for the λ vector with or without a cloned DNA fragment should be attained. Obviously the presence of all mutations contributing to this high degree of biological containment must be verified periodically by appropriate tests. Laboratory tests should be performed with the bacterial host to measure all possible routes of escape such as the frequency of lysogen formation, the frequency of plasmid formation and the survival of the lysogen or carrier bacterium. Similarly, the potential for perpetuation of a cloned DNA fragment carried by infectious phage particles can be tested by challenging typical wild-type *E. coli* strains or a λ-sensitive nonpermissive laboratory K–12 strain, especially one lysogenic for a lambdoid phage.

In view of the fact that accurate assessment of the probabilities for escape of infections λ-grown on r^- m^- Su$^+$ hosts is dependent upon the frequencies of r^-, Su$^+$, and λ-sensitive strains in nature, investigators need to screen *E. coli* strains for these properties. These data will also be useful in predicting frequencies of successful escape of plasmid cloning vectors harbored in r^- m^- Su$^+$ strains.

When any investigator has obtained data on the level of containment provided by a proposed EK2 system, these should be reported as rapidly as possible to permit general awareness and evaluation of the safety features of the new system. Investigators are also encouraged to make such new safer cloning systems generally available to other scientists. NIH will take appropriate steps to aid in the distribution of these safer vectors and hosts.

EK3 host-vectors. These are EK2 systems for which the specified containment shown by laboratory tests has been independently confirmed by appropriate tests in animals, including humans or primates, and in other relevant environments in order to provide additional data to validate the levels of containment afforded by the EK2 host-vector systems. Evaluation of the effects of individual or combinations of mutations contributing to the biological containment should be performed as a means to confirm the degree of safety provided and to further advance the technology of developing even safer vectors and hosts. For the time being, no host-vector system will be considered to be a bona fide EK3 host-vector system, until it is so certified by the NIH Recombinant DNA Molecule Program Advisory Committee.

2. Classification of experiments using the E. coli K–12 containment systems

In the following classification of containment criteria for different kinds of recombinant DNAs, the stated levels of physical and biological containment are minimums. Higher levels of biological containment (EK3 > EK2 > EK1) are to be used if they are available and are equally appropriate for the purposes of the experiment.

(a) Shotgun Experiments. These experiments involve the production of recombinant DNAs between the vector and the total DNA or (preferably) any partially purified fraction thereof from the specified cellular source.

(i) Eukaryotic DNA recombinants —Primates. P3 physical containment + an EK3 host-vector, or P4 physical containment + an EK2 host-vector, except for DNA from uncontaminated embryonic tissue or primary tissue cultures therefrom, and germ-line cells for which P3 physical containment + an EK2 host-vector can be used. The basis for the lower estimated hazard in the case of DNA from the latter tissues (if freed of adult tissue) is their relative freedom from horizontally acquired adventitious viruses.

Other mammals. P3 physical containment + an EK2 host-vector.

Birds. P3 physical containment + an EK2 host-vector.

Cold-blooded vertebrates. P2 physical containment + an EK2 host-vector except for embryonic or germ-line DNA which require P2 physical containment + an EK1 host-vector. If the eukaryote is known to produce a potent toxin, the containment shall be increased to P3 + EK2.

Other cold-blooded animals and lower eukaryotes. This large class of eukaryotes is divided into the following two groups:

(1) Species that are known to produce a potent toxin or are known pathogens (i.e., an agent listed in Class 2 of ref. 5 or a plant pathogen) or are known to carry such pathogenic agents must use P3 physical containment + an EK2 host-vector. Any species that has a demonstrated capacity for carrying particular pathogenic agents is included in this group unless it has been shown that those organisms used as the source of DNA do not contain these agents; in this case they may be placed in the second group.

(2) The remainder of the species in this class can use P2 + EK1. However, any insect in this group should have been grown under laboratory conditions for at least 10 generations prior to its use as a source of DNA.

Plants. P2 physical containment + an EK1 host-vector. If the plant carries a known pathogenic agent or makes a product known to be dangerous to any species, the containment must be raised to P3 physical containment + an EK2 host-vector.

(ii) Prokaryotic DNA recombinants — Prokaryotes that exchange genetic information with E. coli.[2] The level of physical containment is directly determined by the rule of the most dangerous component (see introduction to Section III). Thus P1 conditions can be used for DNAs from those bacteria in Class 1 of ref. 5 ("Agents of no or minimal hazard....") which naturally exchange genes with E. coli; and P2 conditions should be used for such bacteria if they fall in Class 2 of ref. 5 ("Agents of ordinary potential hazard...."), or are plant pathogens or symbionts. EK1 host-vectors can be used for all experiments requiring only P1 physical containment; in fact, experiments in this category can be performed with E. coli K–12 vectors exhibiting a lesser containment (e.g., conjugative plasmids) than EK1 vectors. Experiments with DNA from species requiring P2 physical containment which are of low pathogenicity (for example, enteropathogenic Escherichia coli, Salmonella typhimurium, and Klebsiella pneumoniae) can use EK1 host-vectors, but those of moderate pathogenicity (for example, Salmonella typhi, Shigella dysenteriae type I, and Vibrio cholerae) must use EK2 host-vectors.[3] A specific example of an experiment with a plant pathogen requiring P2 physical containment + an EK2 host-vector would be cloning the tumor gene of Agrobacterium tumefaciens.

Prokaryotes that do not exchange genetic information with E. coli. The minimum containment conditions for this class consist of P2 physical containment + an EK2 host-vector or P3 physical containment + an EK1 host-vector, and apply when the risk that the recombinant DNAs will increase the pathogenicity or ecological potential of the host is judged to be minimal. Experiments with DNAs from pathogenic species (Class 2 ref. 5 plus plant pathogens) must use P3 + EK2.

(iii) Characterized clones of DNA recombinants derived from shotgun experiments. When a cloned DNA recombinant has been rigorously characterized[4] and there is sufficient evidence that it is free of harmful genes,[4] then experiments involving this recombinant DNA can be carried out under P1 + EK1 conditions if the inserted DNA is from a species that exchange genes with E. coli, and under P2 + EK1 conditions if not.

(b) Purified cellular DNAs other than plasmids, bacteriophages, and other viruses. The formation of DNA recombinants from cellular DNAs that have been enriched[5] by physical and chemical techniques (i.e., not by cloning) and which are free of harmful genes can be carried out under lower containment conditions than used for the corresponding shotgun experiment. In general, the containment can be decreased one step in physical containment (P4→P3→P2→P1) while maintaining the biological containment specified for the shotgun experiment or one step in biological containment (EK3→EK2→EK1) while maintaining the specified physical containment — provided that the new condition is not less than that specified above for characterized clones from shotgun experiments (Section (a) — iii).

(c) Plasmids, bacteriophages, and other viruses. Recombinants formed between EK-type vectors and other plasmid or virus DNAs have in common the potential for acting as double vectors because of the replication functions in these DNAs. The containment conditions given below apply only to propagation of the DNA recombinants in E. coli K– 12 hosts. They do not apply to other hosts where they may be able to replicate as a result of functions provided by the DNA inserted into the EK vectors. These are considered under other host-vector systems.

(i) Animal viruses. P4 + EK2 or P3 + EK3 shall be used to isolate DNA recombinants that include all or part of the genome of an animal virus. This recommendation applies not only to experiments of the "shotgun" type but also to those involving partially characterized subgenomic segments of viral DNAs (for example, the genome of defective viruses, DNA fragments isolated after treatment of viral genomes with restriction enzymes, etc). When cloned recombinants have been shown by suitable biochemical and biological tests to be free of harmful regions, they can be handled in P3 + EK2 conditions. In the case of DNA viruses, harmless regions include the late region of the genome; in the case of DNA copies of RNA viruses, they might include the genes coding for capsid proteins or envelope proteins.

(ii) Plant viruses. P3 + EK1 or P2 + EK2 conditions shall be used to form DNA recombinants that include all or part of the genome of a plant virus.

(iii) Eukaryotic organelle DNAs. The containment conditions given below apply only when the organelle DNA has been purified[6] from isolated organelles. Mitochondrial DNA from primates: P3 + EK1 or P2 + EK2. Mitochondrial or chloroplast DNA from other eukaryotes: P2 + EK1. Otherwise, the conditions given under shotgun experiments apply.

(iv) Prokaryotic plasmid and phage DNAs — Plasmids and phage from hosts that exchange genetic information with E. coli. Experiments with DNA recombinants formed from plasmids or phage genomes that have not been characterized with regard to presence of harmful genes or are known to contribute significantly to the pathogenicity of their normal hosts must use the containment conditions specified for shotgun experiments with DNAs from the respective host. If the DNA recombinants are formed from plasmids or phage that are known not to contain harmful genes, or from purified[6] and characterized plas-

mid or phage DNA segments known not to contain harmful genes, the experiments can be performed with P1 physical containment + an EK1 host-vector.

Plasmids and phage from hosts that do not exchange genetic information with E. coli. The rules for shotgun experiments with DNA from the host apply to their plasmids or phages. The minimum containment conditions for this category (P2 + EK2, or P3 + EK1) can be used for plasmid and phage, or for purified[6] and characterized segments of plasmid and phage DNAs, when the risk that the recombinant DNAs will increase the pathogenicity or ecological potential of the host is judged to be minimal.

NOTE. — Where applicable, cDNAs (i.e., complementary DNAs) synthesized in vitro from cellular or viral RNAs are included within each of the above classifications. For example, cDNAs formed from cellular RNAs that are not purified and characterized are included under (a), shotgun experiments; cDNAs formed from purified and characterized RNAs are included under (b); cDNAs formed from viral RNAs are included under (c); etc.

3. Experiments with other prokaryotic host-vectors Other prokaryotic host-vector systems are at the speculative, planning, or developmental stage, and consequently do not warrant detailed treatment here at this time. However, the containment criteria for different types of DNA recombinants formed with E. coli K– 12 host-vectors can, with the aid of some general principles given here, serve as a guide for containment conditions with other host-vectors when appropriate adjustment is made for their different habitats and characteristics. The newly developed host-vector systems should offer some distinct advantage over the E. coli K– 12 host-vectors — for instance, thermophilic organisms or other host-vectors whose major habitats do not include humans and/or economically important animals and plants. In general, the strain of any prokaryotic species used as the host is to conform to the definition of Class 1 etiologic agents given in ref. 5 (i.e., "Agents of no or minimal hazard. . . ."), and the plasmid or phage vector should not make the host more hazardous. Appendix A gives a detailed discussion of the B. subtilis system, the most promising alternative to date.

At the initial stage, the host-vector must exhibit at least a moderate level of biological containment comparable to EK1 systems, and should be capable of modification to obtain high levels of containment comparable to EK2 and EK3. The type of confirmation test(s) required to move a host-vector from an EK2-type classification to an EK3-type will clearly depend upon the preponderant habitat of the host-vector. For example, if the unmodified host-vector propagates mostly in, on, or around higher plants, but not appreciably in warm-blooded animals, modification should be designed to reduce the probability that the host-vector can escape to and propagate in, on, or around such plants, or transmit recombinant DNA to other bacterial hosts that are able to occupy these ecological niches, and it is these lower probabilities which must be confirmed. The following principles are to be followed in using the containment criteria given for experiments with *E. coli* K–12 host-vectors as a guide for other prokaryotic systems. Experiments with DNA from prokaryotes (and their plasmids or viruses) are classified according to whether the prokaryote in question exchanges genetic information with the host-vector or not, and the containment conditions given for these two classes with *E. coli* K–12 host-vectors applied. Experiments with recombinants between plasmid or phage vectors and DNA that extends the range of resistance of the recipient species to therapeutically useful drugs must use P3 physical containment + a host-vector comparable to EK1 or P2 physical containment + a host-vector comparable to EK2. Transfer of recombinant DNA to plant pathogens can be made safer by using nonreverting, doubly auxothrophic, non-pathogenic variants. Experiments using a plant pathogen that affects an element of the local flora will require more stringent containment than if carried out in areas where the host plant is not common.

Experiments with DNAs from eukaryotes (and their plasmids or viruses) can also follow the criteria for the corresponding experiments with *E. coli* K–12 vectors if the major habitats of the given host-vector overlap those of *E. coli*. If the host-vector has a major habitat that does not overlap those of *E. coli* (e.g., root nodules in plants), then the containment conditions for some eukaryotic recombinant DNAs need to be increased (for instance, higher plants and their viruses in the preceding example), while others can be reduced.

4. Experiments with eukaryotic host-vectors

(*a*) *Animal host-vector systems.* Because host cell lines generally have little if any capacity for propagation outside the laboratory, the primary focus for containment is the vector, although cells should also be derived from cultures expected to be of minimal hazard. Given good microbiological practices, the most likely mode of escape of recombinant DNAs from a physically contained laboratory is carriage by humans; thus vectors should be chosen that have little or no ability to replicate in human cells. To be used as a vector in a eukaryotic host, a DNA molecule needs to display all of the following properties:

(1) It shall not consist of the whole genome of any agent that is infectious for humans or that replicates to a significant extent in human cells in tissue culture.

(2) Its functional anatomy should be known — that is, there should be a clear idea of the location within the molecule of:

(a) The sites at which DNA synthesis originates and terminates,

(b) The sites that are cleaved by restriction endonucleases,

(c) The template regions for the major gene products.

(3) It should be well studied genetically. It is desirable that mutants be available in adequate number and variety, and that quantitative studies of recombination have been performed.

(4) The recombinant must be defective, that is, its propagation as a virus is dependent upon the presence of a complementing helper genome. This helper should either (a) be integrated into the genome of a stable line of host cells (a situation that would effectively limit the growth of the vector to that particular cell line) or (b) consist of a defective genome or an appropriate conditional lethal mutant virus (in which case the experiments would be done under non-permissive conditions), making vector and helper dependent upon each other for propagation. However, if none of these is available, the use of a non-defective genome as helper would be acceptable.

Currently only two viral DNAs can be considered as meeting these requirements: these are the genomes of polyoma virus and SV40.

Of these, polyoma virus is highly to be preferred. SV40 is known to propagate in human cells, both *in vivo* and *in vitro,* and to infect laboratory personnel, as evidenced by the frequency of their conversion to producing SV40 antibodies. Also, SV40 and related viruses have been found in association with certain human neurological and malignant diseases. SV40 shares many properties, and gives complementation, with the common human papova viruses. By contrast, there is no evidence that polyoma infects humans, nor does it replicate to any significant extent in human cells *in vitro.* However, this system still needs to be studied more extensively. Appendix B gives further details and documentation.

Taking account of all these factors:

(1) *Polyoma Virus.* (*a*) Recombinant DNA molecules consisting of defective polyoma virus genomes plus DNA sequences of any non-pathogenic organism, including Class 1 viruses (5), can be propagated in or used to transform cultured cells. P3 conditions are required. Appropriate helper virus can be used if needed. Whenever there is a choice, it is urged that mouse cells, derived preferably from embryos, be used as the source of eukaryotic DNA. Polyoma virus is a mouse virus and recombinant DNA molecules containing both viral and cellular sequences are

already known to be present in virus stocks grown at a high multiplicity. Thus, recombinants formed *in vitro* between polyoma virus DNA and mouse DNA are presumably not novel from an evolutionary point of view.

(*b*) Such experiments are to be done under P4 conditions if the recombinant DNA contains segments of the genomes of Class 2 animal viruses (5). Once it has been shown by suitable biochemical and biological tests that the cloned recombinant contains only harmless regions of the viral genome (see Section IIIB-2-c-i) and that the host range of the polyoma virus vector has not been altered, experiments can be continued under P3 conditions.

(2) *SV40 Virus.* (*a*) Defective SV40 genomes, with appropriate helper, can be used as a vector for recombinant DNA molecules containing sequences of any non-pathogenic organism or Class I virus (5), (i.e., a shotgun type experiment). P4 conditions are required. Established lines of cultured cells should be used.

(*b*) Such experiments are to be carried out in P3 (or P4) conditions if the non-SV40 DNA segment is (a) a purified[6] segment of prokaryotic DNA lacking toxigenic genes, or (b) a segment of eukaryotic DNA whose function has been established, which does not code for a toxic product, and which has been previously cloned in a prokaryotic host-vector system. It shall be confirmed that the defective virus-helper virus system does not replicate significantly more efficiently in human cells in tissue culture than does SV40, following infection at a multiplicity of infection of one or more helper SV40 viruses per cell.

(*c*) A recombinant DNA molecule consisting of defective SV40 DNA lacking substantial segments of the late region, plus DNA from non-pathogenic organisms or Class I viruses (5), can be propagated as an autonomous cellular element in established lines of cells under P3 conditions provided that there is no exogenous or endogenous helper, and that it is demonstrated that *no* infectious virus particles are being produced. Until this has been demonstrated, the appropriate containment conditions specified in *2. a.* and *2. b.* shall be used.

(*d*) Recombinant DNA molecules consisting of defective SV40 DNA and sequences from non-pathogenic prokaryotic or eukaryotic organisms or Class I viruses (5) can be used to transform established lines of non-permissive cells under P3 conditions. It must be demonstrated that no infectious virus particles are being produced; rescue of SV40 from such transformed cells by co-cultivation or transfection techniques must be carried out in P4 conditions.

(3) Efforts are to be made to ensure that all cell lines are free of virus particles and mycoplasma.

Since SV40 and polyoma are limited in their scope to act as vectors, chiefly because the amount of foreign DNA that the normal virions can carry probably cannot exceed 2×10^6 daltons, the development of systems in which recombinants can be cloned and propagated purely in the form of DNA, rather than in the coats of infectious agents is necessary. Plasmid forms of viral genomes or organelle DNA need to be explored as possible cloning vehicles in eukaryotic cells.

(*b*) *Plant host-vector systems.* For cells in tissue cultures, seedlings, or plant parts (e.g. tubers, stems, fruits, and detached leaves) or whole mature plants of small species (e.g., *Arabidopsis*) the P1 – P4 containment conditions that we have specified previously are relevant concepts. However, work with most plants poses additional problems. The greenhouse facilities accompanying P2 laboratory physical containment conditions can be provided by: (i) Insect-proof greenhouses, (ii) appropriate sterilization of contaminated plants, pots, soil, and runoff water, and (iii) adoption of the other standard practices for microbiological work. P3 physical containment can be sufficiently approximated by confining the operations with whole plants to growth chambers like those used for work with radioactive isotopes, provided that (i) such chambers are modified to produce a negative pressure environment with the exhaust air appropriately filtered, (ii) that other operations with infectious materials are carried out under the specified P3 conditions, and (iii) to guard against inadvertent insect transmission of recombinant-DNA, growth chambers are to be routinely fumigated and only used in insect proof rooms. The P2 and P3 conditions specified earlier are therefore extended to include these cases for work on higher plants.

The host cells for experiments on recombinant DNAs may be cells in culture, in seedling or plant parts. Whole plants or plant parts that cannot be adequately contained shall not be used as hosts for shotgun experiments at this time, and attempts to infect whole plants with recombinant DNA shall not be initiated until the effects on host cells in culture, seedlings or plant parts have been thoroughly studied.

Organelle or plasmid DNAs or DNAs of viruses of restricted host range may be used as vectors. In general, similar criteria for selecting host-vectors to those given in the preceding section on animal systems are to apply to plant systems.

DNA recombinants formed between the initial moderately contained vectors and DNA from cells of species in which the vector DNA can replicate, require P2 physical containment. However, if the source of the DNA is itself pathogenic or known to carry pathogenic agents, or to produce products dangerous to plants, or if the vector is an unmodified virus of unrestricted host range, the experiments shall be carried out under P3 conditions.

Experiments on recombinant DNAs formed between the above vectors and DNAs from other species can also be carried out under P2 if that DNA has been purified[6] and determined not to contain harmful genes. Otherwise, the experiments shall be carried out under P3 conditions if the source of the inserted DNA is not itself a pathogen, or known to carry such pathogenic agents, or to produce harmful products — and under P4 conditions if these conditions are not met.

The development and use of host-vector systems that exhibit a high level of biological containment permit a decreae of one step in the physical containment specified above ($P4 \rightarrow P3 \rightarrow P2P1$).

(c) *Fungal or similar lower eukaryotic host-vector systems.* The containment criteria for experiments on recombinant DNAs using these host-vectors most closely resemble those for pro-karyotes, rather than those for the preceding eukaryotes, in that the host cells usually exhibit a capacity for dissemination outside the laboratory that is similar to that for bacteria. We therefore consider that the containment guidelines given for experiments with *E. coli* K–12 and other pro-karyotic host-vectors (Sections IIIB–1 and – 2, respectively) provide adequate direction for experiments with these lower eukaryotic host-vectors. This is particularly true at this time since the development of these host-vectors is presently in the speculative stage.

IV. ROLES AND RESPONSIBILITIES

Safety in research involving recombinant DNA molecules depends upon how the research team applies these guidelines. Motivation and critical judgment are necessary, in addition to specific safety knowledge, to ensure protection of personnel, the public, and the environment.

The guidelines given here are to help the principal investigator determine the nature of the safeguards that should be implemented. These guidelines will be incomplete in some respects because all conceivable experiments with recombinant DNAs cannot now be anticipated. Therefore, they cannot substitute for the investigator's own knowledgeable and discriminating evaluation. Whenever this evaluation calls for an increase in containment over that indicated in the guidelines, the investigator has a responsibility to institute such an increase. In contrast, the containment conditions called for in the guidelines should not be decreased without review and approval at the institutional and NIH levels.

The following roles and responsibilities define an administrative framework in which safety is an essential and integrated function of research involving recombinant DNA molecules.

A. PRINCIPAL INVESTIGATOR

The principal investigator has the primary responsibility for: (i) Determining the real and potential biohazards of the proposed research, (ii) determining the appropriate level of biological and physical containment, (iii) selecting the microbiological practices and laboratory techniques for handling recombinant DNA materials, (iv) preparing procedures for dealing with accidental spills and overt personnel contamination, (v) determining the applicability of various precautionary medical practices, serological monitoring, and immunization, when available, (vi) securing approval of the proposed research prior to initiation of work, (vii) submitting information on purported EK2 and EK3 systems to the NIH Recombinant DNA Molecule Program Advisory Committee and making the strains available to others, (viii) reporting to the institutional biohazards committee and the NIH Office of Recombinant DNA Activities new information bearing on the guidelines, such as technical information relating to hazards and new safety procedures or innovations, (ix) applying for approval from the NIH Recombinant DNA Molecule Program Advisory Committee for large scale experiments with recombinant DNAs known to make harmful products (i.e., more than 10 liters of culture), and (x) applying to NIH for approval to lower containment levels when a cloned DNA recombinant derived from a shotgun experiment has been rigorously characterized and there is sufficient evidence that it is free of harmful genes.

Before work is begun, the principal investigator is responsible for: (i) Making available to program and support staff copies of these portions of the approved grant application that describe the biohazards and the precautions to be taken, (ii) advising the program and support staff of the nature and assessment of the real and potential biohazards, (iii) instructing and training this staff in the practices and techniques required to ensure safety, and in the procedures for dealing with accidentally created biohazards, and (iv) informing the staff of the reasons and provisions for any advised or requested precautionary medical practices, vaccinations, or serum collection.

During the conduct of the research, the principal investigator is responsible for: (i) Supervising the safety performance of the staff to ensure that the required safety practices and techniques are employed, (ii) investigating and reporting in writing to the NIH Office of Recombinant DNA Activities and the institutional biohazards committee any serious or extended illness of a worker or any accident that results in (a) inoculation of recombinant DNA materials through cutaneous penetration, (b) ingestion of recombinant DNA materials, (c) probable inhalation of recombinant DNA materials following gross aerosolization, or (d) any incident causing serious exposure to personnel or danger of environmental contamination, (iii) investigating and reporting in writing to the NIH Office of Recombinant DNA Activities and the institutional biohazards committee any problems pertaining to operation and implementation of biological and physical containment safety practices and procedures, or equipment or facility failure, (iv) correcting work errors and conditions that may result in the release of recombinant DNA materials, and (v) ensuring the integrity of the physical containment (e.g., biological safety cabinets) and the biological containment (e.g., genotypic and phenotypic characteristics, purity, etc.).

B. INSTITUTION

Since in almost all cases, NIH grants are made to institutions rather than to individuals, all the responsibilities of the principal investigator listed above are the responsibilities of the institution under the grant, fulfilled on its behalf by the principal investigator. In addition, the institution is responsible for establishing an institutional biohazards committee[7] to: (i) Advise the institution on policies, (ii) create and maintain a central reference file and library of catalogs, books, articles, newsletters, and other communications as a source of advice and reference regarding, for example, the availability and level of biological containment for various host-vector systems, suitable training of personnel and data on the potential biohazards associated with certain recombinant DNAs, (iii) develop a safety and operations manual for any P4 facility maintained by the institution and used in support of recombinant DNA research, (iv) certify to the NIH on applications for research support and annually thereafter, that facilities, procedures, and practices and the training and expertise of the personnel involved have been reviewed and approved by the institutional biohazards committee.

The biohazards committee must be sufficiently qualified through the experience and expertise of its membership and the diversity of its membership to ensure respect for its advice and counsel. Its membership should include individuals from the institution or consultants, selected so as to provide a diversity of disciplines relevant to recombinant DNA technology, biological safety, and engineering. In addition to possessing the professional competence necessary to assess and review specific activities and facilities, the committee should possess or have available to it, the competence to determine the acceptability of its findings in terms of applicable laws, regulations, standards of practices, community attitudes, and health and environmental considerations. Minutes of the meetings should be kept and made available for public inspection. The institution is responsible for reporting names of and relevant background information on the members of its biohazards committee to the NIH.

C. NIH INITIAL REVIEW GROUPS (STUDY SECTIONS)

The NIH Study Sections, in addition to reviewing the scientific merit of each grant application involving recombinant DNA molecules, are responsible for: (i) Making an independent evaluation of the real and potential biohazards of the proposed research on the basis of these guidelines, (ii) determining whether the proposed physical containment safeguards certified by the institutional biohazards committee are appropriate for control of these biohazards, (iii) determining whether the proposed biological containment safeguards are appropriate, (iv) referring to the NIH Recombinant DNA Molecule Program Advisory Committee or the NIH Office of Recombinant DNA Activities those problems pertaining to assessment of biohazards or safeguard determination that cannot be resolved by the Study Sections.

The membership of the Study Sections will be selected in the usual manner. Biological safety expertise, however, will be available to the Study Sections for consultation and guidance.

D. NIH RECOMBINANT DNA MOLECULE PROGRAM ADVISORY COMMITTEE

The Recombinant DNA Molecule Program Advisory Committee advises the Secretary, Department of Health, Education, and Welfare, the Assistant Secretary for Health, Department of Health, Education, and Welfare, and the Director, National Institutes of Health, on a program for the evaluation of potential biological and ecological hazards of recombinant DNAs (molecules resulting from different segments of DNA that have been joined together in cell-free systems, and which have the capacity to infect and replicate in some host cell, either autonomously or as an integrated part of their host's genome), on the development of procedures which are designed to prevent the spread of such molecules within human and other populations, and on guidelines to be followed by investigators working with potentially hazardous recombinants.

The NIH Recombinant DNA Molecule Program Advisory Committee has responsibility for: (i) Revising and updating guidelines to be followed by investigators working with DNA recombinants, (ii) for the time being, receiving information on purported EK2 and EK3 systems and evaluating and certifying that host-vector systems meet EK2 or EK3 criteria, (iii) resolving questions concerning potential biohazard and adequacy of containment capability if NIH staff or NIH Initial Review Group so request, and (iv) reviewing and approving large scale experiments with recombinant DNAs known to make harmful products (e.g., more than 10 liters of culture).

E. NIH STAFF

NIH Staff has responsibility for: (i) Assuring that no NIH grants or contracts are awarded for DNA recombinant research unless they (a) conform to these guidelines, (b) have been properly reviewed and recommended for approval, and (c) include a properly executed Memorandum of Understanding and Agreement, (ii) reviewing and responding to questions or problems or reports submitted by institutional biohazards committees or principal investigators, and disseminating findings, as appropriate, (iii) receiving and reviewing applications for approval to lower containment levels when a cloned DNA recombinant derived from a shotgun experiment has been rigorously characterized and there is sufficient evidence that it is free of harmful genes, (iv) referring items covered under (ii) and (iii) above to the NIH Recombinant DNA Molecule Program Advisory Committee, as deemed necessary, and (v) performing site inspections of all P4 physical containment facilities, engaged in DNA recombinant research, and of other facilities as deemed necessary.

V. FOOTNOTES

[1] Biological Safety Cabinets referred to in this section are classified as *Class I, Class II* or *Class III* cabinets. A *Class I* cabinet is a ventilated cabinet for personnel protection having an inward flow of air away from the operator. The exhaust air from this cabinet is filtered through a high efficiency or high efficiency particulate air (HEPA) filter before being discharged to the outside atmosphere. This cabinet is used in three operational modes; (1) with an 8 inch high full width open front, (2) with an installed front closure panel (having four eight inch diameter openings) without gloves, and (3) with an installed front closure panel equipped with arm length rubber gloves. The face velocity of the inward flow of air through the full width open front is 75 feet per minute or greater. A *Class II* cabinet is a ventilated cabinet for personnel and product protection having an open front with inward air flow for personnel protection, and HEPA filtered mass recirculated air flow for product protection. The cabinet exhaust air is filtered through a HEPA filter. The face velocity of the inward flow of air through the full width open front is 75 feet per minute or greater. Design and performance specifications for *Class II* cabinets have been adopted by the National Sanitation Foundation, Ann Arbor, Michigan.

A *Class III* cabinet is a closed front ventilated cabinet of gas tight construction which provides the highest level of personnel protection of all Biohazard Safety Cabinets. The interior of the cabinet is protected from contaminants exterior to the cabinet. The cabinet is fitted with arm length rubber gloves and is operated under a negative pressure of at least 0.5 inches water gauge. All supply air is filtered through HEPA filters. Exhaust air is filtered through HEPA filters or incinerated before being discharged to the outside environment.

[2] Defined as observable under optimal laboratory conditions by transformation, transduction, phage infection and/or conjugation with transfer of phage, plasmid and/or chromosomal genetic information.

[3] The bacteria which constitute Class 2 of ref. 5 ("Agents of ordinary potential hazard....") represent a broad spectrum of etiologic agents which possess different levels of virulence and degrees of communicability. We think it appropriate for our specific purpose to further subdivide the agents of Class 2 into those which we believe to be of relatively low pathogenicity and those which are moderately pathogenic. The several specific examples given may suffice to illustrate the principle.

[4] The terms "characterized" and "free of harmful genes" are unavoidably vague. But, in this instance, before containment conditions lower than the ones used to clone the DNA can be adopted, the investigator must obtain approval from the National Institutes of Health. Such approval would be contingent upon data concerning: (a) the absence of potentially harmful genes (e.g., sequences contained in indigenous tumor viruses or which code for toxic substances), (b) the relation between the recovered and desired segment (e.g., hybridization and restriction endonuclease fragmentation analysis where applicable), and (c) maintenance of the biological properties of the vector.

[5] A DNA preparation is defined as enriched if the desired DNA represents at least 99% (w/w) of the total DNA in the preparation. The reason for lowering the containment level when this degree of enrichment has been obtained is based on the fact that the total number of clones that must be examined to obtain the desired clone is markedly reduced. Thus, the probability of cloning a harmful gene could, for example, be reduced by more than 10^5-fold when a nonrepetitive gene from mammals was being sought. Furthermore, the level of purity specified here makes it easier to establish that the desired DNA does not contain harmful genes.

[6] The DNA preparation is defined as purified if the desired DNA represents at least 99% (w/w) of the total DNA in the preparation, provided that it was verified by more than one procedure.

[7] In special circumstances, in consultation with the NIH Office of Recombinant DNA Activities, an area biohazards committee may be formed, composed of members from the institution and/or other organizations beyond its own staff, as an alternative when additional expertise outside the institution is needed for the indicated reviews.

VI. REFERENCES

(1) Berg, P., D. Baltimore, H. W. Boyer, S. N. Cohen, R. W. Davis, D. S. Hogness, D. Nathans, R. O. Roblin, J. D. Watson, S. Weissman, and N. D. Zinder (1974). *Potential Biohazards of Recombinant DNA Molecules.* Science *185,303.*

(2) Advisory Board for the Research Councils. *Report of a Working Party on the Experimental Manipulation of the Genetic Composition of Micro-Organisms.* Presented to Parliament by the Secretary of State for Education and Science by Command of Her Majesty. January, 1975. London: Her Majesty's Stationery Office, 1975.

(3) Berg, P., D. Baltimore, S. Brenner, R. O. Roblin and M. F. Singer (1975). *Summary Statement of the Asilomar Conference on Recombinant DNA Molecules.* Science *188,* 991; Nature *225,* 442; Proc. Nat. Acad. Sci. *72,* 1981.

(4) *Laboratory Safety at the Center for Disease Control* (Sept., 1974). U.S. Department of Health, Education, and Welfare Publication No. CDC 75–8118.

(5) *Classification of Etiologic Agents on the Basis of Hazard.* (4th Edition, July, 1974). U.S. Department of Health, Education, and Welfare. Public Health Service. Center for Disease Control, Office of Biosafety, Atlanta, Georgia 30333.

(6) *National Cancer Institute Safety Standards for Research Involving Oncogenic Viruses* (Oct., 1974). U.S. Department of Health, Education, and Welfare Publication No. (NIH) 75–790.

(7) *National Institutes of Health Biohazards Safety Guide* (1974). U.S. Department of Health, Education, and Welfare. Public Health Service, National Institutes of Health. U.S. Government Printing Office Stock No. 1740–00383.

(8) *Biohazards in Biological Research* (1973). A. Hellman, M. N. Oxman and R. Pollack (ed.). Cold Spring Harbor Laboratory.

(9) *Handbook of Laboratory Safety* (1971; 2nd Edition). N. V. Steere (ed.). The Chemical Rubber Co., Cleveland.

(10) Bodily, H. L. (1970). *General Administration of the Laboratory.* H. L. Bodily, E. L. Updyke and J. O. Masons (eds.), Diagnostic Procedures for Bacterial, Mycotic and Parasitic Infections. American Public Health Association, New York. pp. 11–28.

(11) Darlow, H. M. (1969). *Safety in the Microbiological Laboratory.* In J. R. Norris and D. W. Robbins (ed.), Methods in Microbiology. Academic Press, Inc. New York. pp. 169–204.

(12) *The Prevention of Laboratory Acquired Infection (1974).* C. H. Collins, E. G. Hartley, and R. Pilsworth, Public Health Laboratory Service, Monograph Series No. 6.

(13) Chatigny, M. A. (1961). *Protection Against Infection in the Microbiological Laboratory: Devices and Procedures.* In W. W. Umbreit (ed.). Advances in Applied Microbiology. Academic Press, New York, N.Y. *3:* 131–192.

(14) *Design Criteria for Viral Oncology Research Facilities,* U.S. Department of Health, Education, and Welfare, Public Health Service, National Institutes of Health, DHEW Publication No. (NIH) 75–891, 1975.

(15) Kuehne, R. W. (1973). *Biological Containment Facility for Studying Infectious Disease.* Appl. Microbiol. *26:* 239–243.

(16) Runkle, R. S. and G. B. Phillips. (1969). *Microbial Containment Control Facilities.* Van Nostrand Reinhold, New York.

(17) Chatigny, M. A. and D. I. Clinger (1969). *Contamination Control in Aerobiology.* In R. L. Dimmick and A. B. Akers (eds.). An Introduction to Experimental Aerobiology. John Wiley & Sons, New York, pp. 194–263.

(18) Grunstein, M. and D. S. Hogness (1975). *Colony Hybridization: A Method for the Isolation of Cloned DNAs That Contain a Specific Gene.* Proc. Nat. Acad. Sci. U.S.A. *72,* 3961–3965.

(19) Morrow, J. F., S. N. Cohen, A. C. Y. Chang, H. W. Boyer, H. M. Goodman and R. B. Helling (1974). *Replication and Transcription of Eukaryotic DNA in Escherichia coli.* Proc. Nat. Acad. Sci. USA *71,* 1743–1747.

(20) Hershfield, V., H. W. Boyer, C. Yanofsky, M. A. Lovett and D. R. Helinski (1974). *Plasmid Co1E1 as a Molecular Vehicle for Cloning and Amplification of DNA.* Proc. Nat. Acad. Sci. USA *71,* 3455–3459.

(21) Wensink, P. C., D. J. Finnegan, J. E. Donelson, and D. S. Hogness (1974). *A System for Mapping DNA Sequences in the Chromosomes of Drosophila melanogaster.* Cell *3,* 315–325.

(22) Timmis, K. F. Cabello and S. N. Cohen (1974). *Utilization of Two Distinct Modes of Replication by a Hybrid Plasmid Constructed In Vitro from Separate Replicons.* Proc. Nat. Acad. Sci. USA *71,* 4556–4560.

(23) Glover, D. M., R. L. White, D. J. Finnegan and D. S. Hogness (1975). *Characterization of Six Cloned DNAs from Drosophila melanogaster. Including one that Contains the Genes for λRNA.* Cell *5,* 149–155.

(24) Kedes, L. H., A. C. Y. Chang, D. Houseman and S. N. Cohen (1975). *Isolation of Histone Genes from Unfractionated Sea Urchin DNA by Subculture Cloning in E. coli.* Nature *255,* 533.

(25) Tanaka, T. and B. Weisblum (1975). *Construction of a Colicin E1-R Factor Composite Plasmid In Vitro: Means for Amplification of Deoxyribonucleic Acid.* J. Bacteriol. *121,* 354–362.

(26) Tanaka, T., B. Weisblum, M. Schnoss and R. Inman (1975). *Construction and Characterization of a Chimeric Plasmid Composed of DNA from Escherichia coli and Drosophila melanogaster.* Biochemistry *14,* 2064–2072.

(27) Thomas, M., J. R. Cameron and R. W. Davis (1974). *Viable Molecular Hybrids of Bacteriophage Lambda and Eukaryotic DNA.* Proc. Nat. Acad. Sci. USA *71,* 4579–4583.

(28) Murray, N. E. and K. Murray (1974). *Manipulation of Restriction Targets in Phage λ to form Receptor Chromosomes for DNA Fragments.* Nature *251,* 476–481.

(29) Rambach, A. and P. Tiollais (1974). *Bacteriophage λ Having EcoR1 Endonuclease Sites only in the Nonessential Region of the Genome.* Proc. Nat. Acad. Sci. USA *71,* 3927–3930.

(30) Smith, H. W. (1975). *Survival of Orally-Administered Escherichia coli K12 in the Alimentary Tract of Man.* Nature *255,* 500–502.

(31) Anderson, E. S. (1975). *Viability of, and Transfer of a Plasmid from Escherichia coli K12 in the human intestine.* Nature *255,* 502–504.

(32) Falkow, S. (1975). Unpublished experiments quoted in Appendix D of the *Report of the Organizing Committee of the Asilomar Conference on Recombinant DNA Molecules.* (P. Berg, D. Baltimore, S. Brenner, R. O. Robin and M. Singer, eds.) submitted to the National Academy of Sciences.

(33) R. Curtis III, personal communication.

(34) Novick, R. P. and S. I. Morse (1967). *In Vivo* Transmission of Drug Resistance *Factors between Strains of Staphylococcus aureus.* J. Exp. Med. *125,* 45–59.

(35) Anderson, J. D., W. A. Gillespie and M. H. Richmond. (1974). *Chemotherapy and Antibiotic Resistance Transfer between Enterobacteria in the Human Gastrointestinal Tract.* J. Med. Microbiol. *6,* 461–473.

(36) Ronald Davis, personal communication.

(37) K. Murray, personal communication; W. Szybalski, personal communication.

(38) Manly, K. R., E. R. Signer and C. M. Radding (1969).

Nonessential Functions of Bacteriophage λ. Virology *37* 177.

(39) Gottesman, M. E. and R. A. Weisberg (1971). *Prophage Insertion and Excision.* In The Bacteriophage Lambda (A. D. Hershey, ed.). Cold Spring Harbor Laboratory pp. 113–138.

(40) Shimada, K., R. A. Weisberg and M. E. Gottesman (1972). *Prophage Lambda at Unusual Chromosomal Locations: I. Location of the Secondary Attachment Sites and the Properties of the Lysogens.* J. Mol. Biol. *63,* 483–503.

(41) Signer, E. (1969). *Plasmid Formation: A New Mode of Lysogeny by Phage* λ. Nature *223,* 158–160.

(42) Adams, M. H. (1959). *Bacteriophages.* Intersciences Publishers, Inc., New York.

(43) Jacob, F. and E. L. Wollman. (1956). *Sur les Processus de Conjugaison et de Recombinasion chez Escherichia coli. I. L'induction par Conjugaison ou Induction Zygotique.* Ann. Inst. Pasteur *91,* 486–510.

(44) J. S. Parkinson as cited (p. 8) by Hershey, A. D. and W. Dove (1971). *Introduction to Lambda. In:* The Bacteriophage λ. A. D. Hershey, ed. Cold Spring Harbor Laboratory, New York.

Guidelines for the Use of Recombinant DNA Molecule Technology in the City of Cambridge, January 1977

The report of the Cambridge Experimentation Review Board, submitted in January of 1976, has been reprinted here without the letter of transmittal and the appendices.

INTRODUCTION

The Cambridge Experimentation Review Board (CERB) has spent nearly four months studying the controversy over the use of the recombinant DNA technology in the City of Cambridge. The following charge was issued to the Board by the City Manager at the request of the City Council on August 6, 1976.

The broad possibility of the Experimentation Review Board shall be to consider whether research on recombinant DNA which is proposed to be conducted at the P3 level of containment in Cambridge may have any adverse effect on public health within the City, and for this purpose to undertake, among other studies to:

(a) review the "Decision of the Director, National Institutes of Health to Release Guidelines for Research on Recombinant DNA Molecules" dated and released on June 23, 1976;

(b) review but not be limited to the methods of physical and biological containment recommended by the NIH;

(c) review methods for monitoring compliance with applicable procedural safeguards;

(d) review methods for monitoring compliance with safeguards applicable to physical containment;

(e) review procedures for handling accidents (e.g. fire) in recombinant DNA research facilities;

(f) advise the Commissioner of Health and Hospitals on the reviews, findings and recommendations.

Throughout our inquiry we recognized that the controversy over recombinant DNA research involves profound philosophical issues that extend beyond the scope of our charge. The social and ethical implications of genetic research must receive the broadest possible dialogue in our society. That dialogue should address the issue of whether all knowledge is worth pursuing. It should examine whether any particular route to knowledge threatens to transgress upon our precious human liberties. It should raise the issue of technology assessment in relation to long range hazards to our natural and social ecology. Finally, a

152

national dialogue is needed to determine how such policy decisions are resolved in the framework of participatory democracy.

In the several months of testimony, we have come to appreciate the brilliant scientific achievements made in molecular biology and genetics. Recombinant DNA technology promises to contribute to our fundamental knowledge of life processes by providing basic understanding of the function of the gene. The benefits to be derived from this research are uncertain at this time, but the possibility for advancement in clinical medicine as well as in other fields surely exists. While we should not fear to increase our knowledge of the world, to learn more of the miracle of life, we citizens must insist that in the pursuit of knowledge appropriate safeguards be observed by institutions undertaking the research. Knowledge, whether for its own sake or for its potential benefits to humankind, cannot serve as a justification for introducing risks to the public unless an informed citizenry is willing to accept those risks. Decisions regarding the appropriate course between the risks and benefits of potentially dangerous scientific inquiry must not be adjudicated within the inner circles of the scientific establishment. Moreover, the public's awareness of scientific results that have an important impact on society should not depend on crisis situations. Many of the fears over scientific research held by the citizenry result from a lack of understanding about the nature of and the manner in which the research is conducted.

The members of CERB have made a determined effort to assess the risks to the Cambridge community of recombinant DNA research at the P3 level of physical containment. NIH, in issuing its guidelines, sought a balance between "stifling research through excessive regulation and allowing it to continue with sufficient controls." The function of CERB was not to repeat NIH's long and careful deliberation, perhaps one of the most intensive biohazards studies in the history of biology. Our role was to examine the controversy within science. We called upon people from diverse fields to testify. We encouraged skepticism, and in doing so were able to determine the locus of the controversy.

Many of us felt that it was the role of the proponents of the research to justify that *no reasonable likelihood* exists in which the public's health would be compromised if the research is undertaken under the guidelines issued by NIH. We recognized that absolute assurance was an impossible expectation. It was clearly a question of how much assurance was satisfactory to the deliberating body, and in the case of CERB, that body was comprised of citizens with no special interests in promoting the research. The uncertainty we faced was not something fabricated in our community. It was expressed most eloquently by Donald Frederickson, the Director of NIH, when he issued the guidelines:

"In many instances, the views presented to us were contradictory. At present, the hazards may be guessed at, speculated about, or voted upon, but they cannot be known absolutely in the absence of firm experimental data—and unfortunately, the needed data were, more often than not, unavailable."

Our recommendations call for more assurance than was called for by the NIH guidelines. We feel that under our recommendations, a sufficient number of safeguards have been built into the research to protect the public against *any reasonable likelihood* of a biohazard. For *extremely unlikely possibilities,* we have called for additional health monitoring, whereby appropriate personnel are responsible for the detection of hazardous agents, inadvertantly produced, before they are able to threaten the health of the citizens in our community.

We recognize that the controversy over the use of the recombinant DNA technology was brought to the public's attention by a small group of scientists with a deep concern for their fellow citizens and responsibility to their profession. Many of these early critics are now satisfied that the potential hazards of the research are negligible when carried out under the NIH guidelines. There are also those scientists who continue to call for more stringent control over this technology, in many instances, against the majority view of their colleagues and amidst very strained personal relations. To them we owe our gratitude for broadening the context in which the issues are being discussed. The willingness of scientists on both sides of the controversy to share their knowledge with us in our determination to arrive at a reasoned decision has been an inspiration.

CERE has spent over one hundred hours in hearing testimony and carrying out its deliberations. Our decision is as unemotional and as objective as we are capable of offering. It provides a statement of conditions and safeguards that we deem necessary for P3 recombinant DNA research to be carried out in Cambridge. The members of this citizen committee have no association with the biological research in question and no member of the Board has ever had formal ties to the institutions proposing the research, with the exception of one member who has taught in unallied areas at both the institutions in question. Moreover, the City Manager in selecting a group of citizens representing a cross section of the Cambridge community insured that the "empathy factor"—that is, the concern that the institutions proposing the research might lose valuable funds or that qualified researchers would leave in the event of a ban on the research, was never an issue in the deliberations.

In presenting the results of our findings we wish also to express our sincere belief that a predominantly lay citizen group can face a technical scientific matter of general and deep public concern, educate itself appropriately to the task, and reach a fair decision.

SECTION 1
[CONDITIONS FOR
P3 RECOMBINANT DNA RESEARCH
IN CAMBRIDGE]

After reviewing the guidelines issued by the Director of the National Institutes of Health (NIH) for Research Involving Recombinant DNA Molecules (issued June 23, 1976) it is the unanimous judgement of the Cambridge Experimentation Review Board that Recombinant DNA research can be permitted in Cambridge provided that:

The research is undertaken with strict adherence to the NIH Guidelines and in addition to those guidelines the following conditions are met:

I. Institutions proposing recombinant DNA research or proposing to use the recombinant DNA technology shall prepare a manual which contains all procedures relevant to the conduct of said research at all levels of containment and that training in appropriate safeguards and procedures for minimizing potential accidents should be mandatory for all laboratory personnel.

II. The institutional Biohazards Committee mandated by the NIH Guidelines should be broad-based in its composition. It should include members from a variety of disciplines, representation from the bio-technicians staff and at least one community representative unaffiliated with the institution. The community representative should be approved by the Health Policy Board of the City of Cambridge.

III. All experiments undertaken at the P3 level of physical containment shall require an NIH certified host-vector system of at least an EK2 level of biological containment.

IV. Institutions undertaking recombinant DNA experiments shall perform adequate screening to insure the purity of the strain of host organisms used in the experiments and shall test organisms resulting from such experiments for their resistance to commonly used therapeutic antibiotics.

V. As part of the institution's health monitoring responsibilities it shall in good faith make every attempt, subject to the limitation of the available technology, to monitor the survival and escape of the host organism or any component thereof in the laboratory worker. This should include whatever means is available to monitor the intestinal flora of the laboratory worker.

VI. A Cambridge Biohazards Committee (CBC) be established for the purpose of overseeing all recombinant DNA research that is conducted in the City of Cambridge.

A. The CBC shall be composed of the Commissioner of Public Health, the Chairman of the Health Policy Board and a minimum of three members to be appointed by the City Manager.

B. Specific responsibilities of the CBC shall include:

1. Maintaining a relationship with the institutional biohazards committees.

2. Reviewing all proposals for recombinant DNA research to be conducted in the City of Cambridge for compliance with the current NIH guidelines.

3. Developing a procedure for members of institutions where the research is carried on to report to the CBC violations either in technique or established policy.

4. Reviewing reports and recommendations from local institutional biohazards committees.

5. Carrying out site visits to institutional facilities.

6. Modifying these recommendations to reflect future developments in federal guidelines.

7. Seeing that conditions designated as I–V in this section are adhered to.

SECTION 2
[CITY ORDINANCE]

We recommend that a city ordinance be passed to the effect that any recombinant DNA molecule experiments undertaken in the city which are not in strict adherence to the NIH guidelines as supplemented in Section 1 of this report constitute a health hazard to the City of Cambridge.

SECTION 3
[GENERAL RECOMMENDATIONS
TO FEDERAL AUTHORITIES]

We urge that the City Council of Cambridge, on behalf of this Board and the citizenry of the country, make the following recommendations to the Congress:

I. That all uses of recombinant DNA molecule technology fall under uniform federal guidelines and that legislation be enacted in Congress to insure conformity to such guidelines in all sectors, both profit and non-profit, whether such legislation takes a form of licensing or regulation, and that Congress appropriate sufficient funding to adequately enforce compliance with the legislation.

II. That the NIH or other agencies funding recombinant DNA research require institutions to include a health monitoring program as part of their funding proposal and that monies be provided to carry out the monitoring.

III. That a federal registry be established of all workers participating in recombinant DNA research for the purpose of long term epidemiological studies.

IV. That federal initiative be taken to sponsor and fund research to determine the survival and escape of the host organism in the human intestine under laboratory conditions.

SECTION 4
[A DISCUSSION
OF CERB'S REVIEW PROCESS]

In the event that the citizens of Cambridge, the members of the City Council or other interested parties wish to know how the Cambridge Experimentation Review Board carried out its charge to review P3 recombinant DNA research in the City, the final section of this report discusses the review process. In this discussion we include a brief chronology of events, some of the strategies undertaken by the Board for self-education and a description of its deliberation process.

On July 7, 1976, after having held two days of public hearings, the City Council of Cambridge voted a three month "good faith" moratorium on all P3 level recombinant DNA research in the City and called for the establishment of a citizen review board to study the issue.

James L. Sullivan, City Manager of Cambridge released the charge to the newly designated Cambridge Experimentation Review Board (CERB) on August 6, 1976, and issued the guidelines under which that body was to carry out its responsibilities. In addition, eight citizens and the newly appointed acting Commissioner of Health and Hospitals for the City were selected to constitute the Board. Members of the Board were chosen to reflect a cross section of the Cambridge community (see Appendix A). Of the eight citizen Board members, only three had ever met before. Seven of the eight had never had formal ties with either

institution proposing the new research. The one individual who did have some formal ties with the universities has taught courses in structural engineering at both Harvard and M.I.T.

CERB commenced its first meeting August 26, 1976, and continued its hearings until the recommendations of the Board were issued to the Commissioner of Health and Hospitals on December 21, 1976. Meetings were held twice weekly with each session lasting in excess of two hours.

At the September 14th meeting, the Board arrived at a consensus on key policy issues related to the process of its inquiry. Dr. Francis Comunale, initially serving as chairperson, released the chair to the vice chairperson, Daniel Hayes. This decision was made to preclude any ambiguity or conflict of interest in having Dr. Comunale, the then acting Commissioner of Health and Hospitals in the role as chairman of the Board and the person to whom the Board advised on the matter in question. Dr. Comunale thereafter became an ex officio member of the Board. He attended meetings, without a vote, and excluded himself from the final deliberations leading to a decision.

At the same meeting CERB voted to request an extension of the moratorium for an additional three months, on the grounds that we needed the additional time to carry out the full scope of our charge, including a review of the Environmental Impact Statement, which at that time was not complete. The request for an extension of the moratorium was subsequently granted by the City Council and accepted by the institutions affected by the moratorium.

It was agreed that on all decisions undertaken by CERB a consensus would be sought; if consensus could not be reached on an issue, the majority decision would prevail. Moreover, any Board member had the right to poll the entire membership on any issue requiring a vote. If consensus could not be reached on the final recommendation, then minority statements would be permitted in the Board's final report. The members agreed that Thursday meetings would be kept open for the public and the media, while Tuesday sessions would be held in private.

Among the more formidable problems facing this lay citizen board was its self-education. At the outset of the inquiry, the members of CERB were, for the most part, unfamiliar with the concepts, the basic scientific principles and the explanatory models underlying the recombinant DNA technology. The education of the Board members was carried out simultaneously with the inquiry process: We had to decide on the kind of information we would need to reach a decision as well as the kind of people who could provide us with that information.

There were several facets to the Board's information gathering and self-education strategies as exemplified in the following:

*Each Board member was provided with special technical documents on the controversy, including the NIH guidelines, the Environmental Impact Statement, and essays in journals such as *Science.* Along with technical materials, articles that were published in the more popular press and written for a wider readership were distributed to the Board members. As examples, the Board had articles from *Scientific American,* the *New York Times Magazine,* and *National Geographic.*

*A technical assistant to the Board, who had training in the biological sciences, offered help with translating technical concepts. The technical assistant also made available to the Board current articles, news analyses, and essays in leading journals relating to the controversy.

*Spokespeople who appeared before the Board were asked to reduce technical concepts to layman's terms, to present simplified models of bio-chemical events, and to draw upon analogies that helped foster understanding whenever they were available.

*The members of CERB were witness to a forum on the recombinant DNA controversy in which proponents and opponents of the research presented their arguments and responded to questions from the audience.

*Two open line telephone conversations were used to draw testimony from people outside the state. In one of these conversations, the Director of NIH and a panel of experts responded to questions of the Board members.

*In a five hour marathon session, CERB carried out a type of mock courtroom affair. Board members served as a kind of jury, while advocates on both sides of the issue presented their case, were given an opportunity to cross-examine one another, and responded to questions raised by the "citizen jury." This format enabled the Board members to evaluate how well scientists on each side of the controversy responded to the critical issues. Medical researchers and clinicians were also on hand to respond to testimony.

*Board members were taken through laboratories at Harvard and M.I.T. In one case a mock experiment was carried out which exemplified the various stages of the recombinant DNA process. Visiting the laboratories also helped the Board members concretize many of the specifications found in the NIH guidelines relating to physical containment.

Speakers appeared before the Board both on a voluntary basis and at the Board's request. The schedule of speakers called for fair representation of opponents and proponents views, as well as other persons who were called upon to broaden our understanding of the issues. Individuals on each side of the issue were heard from on intermittent weeks.

Some members of CERB visualized the Board as a kind of "citizen jury" whose function it was to review and assess the significance of the recombinant DNA controversy within science. The use of the legal metaphor helped members of the Board clarify for themselves the role of lay citizens in this complex issue. The analogy was of only limited value since Board members functioned in a greater variety of ways than citizens called upon to jury duty. CERB determined the rules of its inquiry, called upon people to testify, listened to the arguments, cross-examined scientists and finally came out with its recommendations.

The use of a "citizen court" in areas of controversy within science that have significant bearing on public welfare is quite new and untested. It encouraged discussions among Board members about where justification rests. At issue was whether the proponents of the research must prove that it is safe beyond all reasonable doubt or whether the opponents must prove that if recombinant DNA research were undertaken there would be significant potential hazards. There was no clear consensus on the issue of who must justify what, and to what degree of satisfaction, however, CERB carried out its inquiry by seeking the strongest positions on both sides of the controversy, while simultaneously looking for weaknesses in the arguments.

Several intensive planning sessions were used to explore CERB's unresolved questions and to draw as wide a range of input from its citizen members as possible. The planning sessions were designed to overcome the factors that inhibit people from expressing their uncertainties. The aim was to eliminate any social hierarchies that could prevent full cooperation and participation from Board members. The success of full cooperation hinged upon the building of confidence for each individual member.

The planning strategy involved first covering the walls of a room with large sheets of paper. Then, a scribe wrote down suggestions from Board members, insuring that each individual completed his/her recommendations or queries before the issues were debated by the entire Board. Finally, the material on the sheets was reduced and synthesized by a technical assistant and sent out to the Board members for discussion at subsequent meetings. This method insures that each citizen member, whatever his/her stand on the controversy, and whatever his/her state of knowledge on the issues, had an unfettered opportunity for self-expression and participation.

Individuals appearing before the Board spent up to three hours discussing the issues and responding to questions. The members of CERB heard over 75 hours of testimony from more than 35 individuals representing both sides of the controversy. In addition, we spent over 25 hours in formal planning and deliberation, as well as countless hours of reviewing related written material before arriving at our decision.

Finally, it is worthwhile noting that despite a considerable heterogeneity in the Board's makeup and differences in how its members initially perceived the controversy, we were able to reach a unanimous decision.

APPENDIX V

Interim Report of the Federal Interagency Committee on Recombinant DNA Research: Suggested Elements for Legislation, March 1977

I. INTRODUCTION

Recent scientific developments in genetics, particularly in the last four years, have culminated in the ability to join together genetic material from different sources in cell-free systems to form recombinant deoxyribonucleic acid (DNA) molecules. DNA is the material that determines hereditary characteristics of all known cells. Recombinant DNA research offers great promise for better understanding and improved treatment of human diseases. Medical advances through use of this technology include the opportunity to explore complicated diseases and the functioning of cells, to better understand a variety of hereditary defects, and possibly in the future, to create microorganisms useful in producing medically important compounds for the treatment and control of disease. Aside from the potential medical benefits, a variety of other applications in science and technology are envisioned. An example is the large-scale production of enzymes for industrial use. Potential benefits in agriculture include the enhancement of nitrogen fixation in certain plants and the biological control of pests, permitting increased food production.

There are risks in this new research area as well as anticipated benefits. A potential hazard, for example, is that the foreign DNA in a microorganism may alter it in unpredictable and undesirable ways. Should the altered microorganism escape from containment, it might infect human beings, animals, or plants, causing disease or modifying the environment. Or the altered bacteria might have a competitive advantage, enhancing their survival in some niche within the ecosystem.

Until the potential risks are better delineated and evaluated in light of developing scientific knowledge, the public should expect such research to be conducted under strict conditions ensuring safety. This was the fundamental principle that guided the Federal Interagency Committee on Recombinant DNA Research in its deliberations — that is, the desire to allow this significant research to continue while simultaneously protecting, as much as humanly possible, man and the environment from effects of potential hazards whose nature is as yet unknown.

158

The Committee formally adopted this interim report by unanimous consent, save for abstentions by the representatives from the Council on Environmental Quality and the Department of Justice.

II. DEVELOPMENT OF THE NIH GUIDELINES ON RECOMBINANT DNA RESEARCH

Approximately three years ago, because of the perceived potential hazards, scientists engaged in this research voluntarily called for a moratorium on certain experiments pending an assessment of risk and the development of appropriate guidelines. These scientists called upon the National Institutes of Health (NIH), of the Department of Health, Education, and Welfare, to create an advisory committee to develop such guidelines. After what NIH considered to be extensive scientific and public review, it released guidelines on June 23, 1976, which established strict conditions for the conduct of NIH-supported research in this area. The NIH Guidelines prohibit certain types of experiments and require special safety conditions for other types. The provisions are designed to afford protection with a wide margin of safety to workers and the environment. The NIH Guidelines were published in the *Federal Register* on July 7, 1976, for public comment.

The NIH also prepared and filed in the *Federal Register* on September 9, 1976, a Draft Environmental Impact Statement on the Guidelines for public comment. The final NIH Environmental Impact Statement will be published shortly. In August 1976 the NIH published a volume containing the transcript of a public hearing held on the Guidelines as well as the correspondence received by the NIH Director on this matter prior to the release of the Guidelines in June.

III. FEDERAL INTERAGENCY COMMITTEE ON RECOMBINANT DNA RESEARCH

The Interagency Committee on Recombinant DNA Research was created to address extension of the NIH Guidelines beyond the NIH to the public and private sectors. The Committee was convened by the Secretary of Health, Education, and Welfare with the approval of the President. Dr. Donald S. Fredrickson, Director of NIH, serves as chairman at the Secretary's request. The Interagency Committee is composed of representatives of Federal Departments and agencies that support or conduct recombinant DNA research, or that may do so in the future, and representatives of Federal Departments and agencies that have present or potential regulatory authority in this area. (The

membership of the Committee is included in Appendix I.) The mandate of the Committee is to

(1) review the nature and scope of Federal- and private-sector activities relating to recombinant DNA research;
(2) determine the extent to which the NIH Guidelines may currently be applied to research in the public and private sectors;
(3) recommend, if appropriate, legislative or executive actions necessary to ensure compliance with the standards set for this research; and
(4) provide for the full communication and necessary exchange of information on recombinant-DNA-research programs and activities throughout the Federal sector.

Two meetings of the Committee were held in November 1976. The first of these, on November 4, was devoted to a review of the development of the NIH Guidelines for Research Involving Recombinant DNA Molecules. The Committee also reviewed activities in other countries on the development of guidelines for this research. Recombinant DNA research is being conducted in a number of countries, including Canada, the United Kingdom, most of Western Europe, the Scandinavian countries, Eastern Europe, the Soviet Union, and Japan.

In many countries appropriate governmental or scientific bodies have reviewed the research and have agreed that it should proceed. Several of the countries have acted to establish guidelines to govern the conduct of this research, including the United Kingdom and Canada. In the United Kingdom a parliamentary committee addressed the issue and indicated that work in this area should continue under appropriate safety conditions. Scientific advisory committees of international organizations, such as the World Health Organization, the International Councils of Scientific Unions, and the European Molecular Biology Organization, have made similar recommendations.

The European Science Foundation, representing member nations from Western Europe and Scandinavia, has recommended to its members that they follow the guidelines of the United Kingdom. These guidelines are, in intent and substance, very similar to those of the National Institutes of Health. The NIH is currently working closely with the United Kingdom and the European Science Foundation to ensure a commonality of standards in the conduct of this research. Thus far, there has been very close cooperation and coordination among the various international and national scientific bodies, with a view to reaching a consensus on safety practices, programs, and procedures.

At the meeting of the Committee held on November 23, the Federal research agencies discussed their activities and possible roles in the implementation of the NIH Guidelines. All Federal research agencies endorsed the Guidelines to govern recombinant DNA research. At present, the NIH, the National Science Foundation, the Veterans Administration, and the U.S. Department of Agriculture are supporting or conducting such research. The NIH has 123 grants in which recombinant DNA research is involved. The National Science Foundation has 52 grants supporting such research in whole or in part. The Veterans Administration has eight projects. The Department of Agriculture and Agricultural Experiment Stations will soon have an estimate of the number of projects in their area. The Department of Defense, the National Aeronautics and Space Administration, and the Energy Research and Development Administration do not at present conduct such research, but all have endorsed the NIH Guidelines to govern future research should it be undertaken.

IV. SUBCOMMITTEE REVIEW OF EXISTING LEGISLATION

At the November 23 meeting of the Interagency Committee, the Federal regulatory agencies also reported on their regulatory functions. Following that review, a special Subcommittee was formed to analyze the relevant statutory authorities for the possible regulation of recombinant DNA research. All regulatory agencies were represented on the Subcommittee, assisted by attorneys from their offices of general counsel. (See Appendix II for the membership of the Subcommittee.) The Subcommittee held meetings on December 13, 1976, and on January 11 and February 8, 1977.

The Subcommittee was charged to determine whether existing legislative authority would permit the regulation of all recombinant DNA research in the United States (whether or not Federally funded) and would include at least the following regulatory requirements:

(1) review of such research by an institutional biohazards committee before it is undertaken,
(2) compliance with physical and biological containment standards and prohibitions in the NIH Guidelines,
(3) registration of such research with a national registry at the time the research is undertaken (subject to appropriate safeguards to protect proprietary interests), and
(4) enforcement of the above requirements through monitoring, inspection, and sanctions.

It was the conclusion of the Subcommittee that present law could permit imposition of some of the above requirements on much recombinant DNA laboratory research, but that no single legal authority or combination of authorities currently exists that would clearly reach all research and other uses of recombinant DNA techniques and meet all the requirements. The complete Subcommittee analysis is included in Appendix III. The Subcommittee, in reaching this conclusion, reviewed the following laws that were deemed most deserving of detailed considerations:

(1) the Occupational Safety and Health Act of 1970 (Public Law 91–596),
(2) the Toxic Substances Control Act (Public Law 94–469),
(3) the Hazardous Materials Transportation Act (Public Law 93–633),
(4) Section 361 of the Public Health Service Act (42 U.S.C. Sec. 264).

The Occupational Safety and Health Act gives the Occupational Safety and Health Administration (OSHA) broad powers to require employers to provide a safe workplace for their employees. The term "employer" in the Act, however, is defined in such a way as to exclude States and their political subdivisions unless the OSHA standards are voluntarily adopted. Twenty-four states have adopted the standards, but twenty-six states are not subject to them. Further, the OSHA standards do not cover self-employed persons. For these reasons it was determined that OSHA at present could not regulate all recombinant DNA research.

The Environmental Protection Agency, under the Toxic Substances Control Act, is directed to control chemicals that may present an "unreasonable risk of injury to the health or the environment." The Subcommittee determined that the materials used in recombinant DNA research would appear to be covered in most cases by the Act's definition of "chemical substance." Section 5 of the Act, however, explicitly exempts registration of chemical substances used in small quantities for the purposes of scientific experimentation or analysis. This represents a most serious deficiency, as the registration of activities was thought to be an essential element of any regulatory effort. Also, in order to meet the specifications of the Act, recombinant DNA research would have to be found to present "an unreasonable risk of injury to health or the environment."

The Hazardous Materials Transportation Act (HMTA) and Section 361 of the Public Health Service (PHS) Act give the Department of Transportation (DOT) and the Center for Disease Control (CDC), respectively, authority to regulate the shipment of hazardous materials in interstate commerce. Both the DOT and the CDC, in implementing these acts with respect to biological products, have essentially aimed at imposing labeling, packaging, and shipping requirements, and were found to be wanting for regulation of all recombinant DNA research.

The Environmental Defense Fund, in November 1976, petitioned the DHEW to regulate recombinant DNA research under Section 361 of the PHS Act. (The petition is included in Appendix IV.) The Subcommittee carefully reviewed this section, which is directed to organisms that are communicable and cause human disease. Thus, under this section, there would have to be a reasonable basis for concluding that the products of all recombinant DNA research may cause human disease and are communicable. Further, Section 361 does not apply to plants, animals, or the general environment. It was the conclusion of the Subcommittee that Section 361 lacked the requisite authority to meet all of the requirements set for the regulation of this research.

The Subcommittee also considered the authority of the CDC to license and control the operation of clinical laboratories under Section 353 of the PHS Act, but this provision was not considered to be applicable to research laboratories.

Other authorities of EPA under the Clean Air Act, the Federal Water Pollution Control Act, and the Resource Conservation and Recovery Act of 1976 were considered briefly and thought only to apply, if at all, to isolated aspects of recombinant DNA research. The authorities of the Food and Drug Administration (FDA) were also reviewed, but it was concluded that recombinant DNA research has not yet reached the stage of commercial application that comes under the FDA's jurisdiction. The regulatory powers of the U.S. Department of Agriculture (USDA) were also reviewed and found applicable solely to nonhuman animals and plants.

In summary, the group concluded that no single legal authority, or combination of authorities, currently exists which would clearly reach all recombinant DNA research in a manner deemed necessary by the Committee. Although there is existing authority that might be broadly interpreted to cover most of the research at issue, it was generally agreed that regulatory actions taken on the basis of any such interpretation would probably be subject to legal challenge.

After completing an analysis of existing legislation, the Subcommittee on February 8, 1977, considered elements which might be included in legislation to regulate recombinant DNA research. The Subcommittee referred the analysis of existing legislation and elements for new legislation to the full Committee at a meeting held on February 25, 1977. The full Committee adopted the report

of the Subcommittee on existing legislation and agreed that new legislation was required.

V. SUGGESTED ELEMENTS FOR LEGISLATION

In considering the elements for legislation, the Committee reviewed Federal, State, and local activities bearing on the regulation of recombinant DNA research. Among congressional proposals reviewed were Senate Bill 621, "The DNA Research Act of 1977," introduced by Senator Dale Bumpers, and the companion measure introduced by Representative Richard L. Ottinger in the House (H.R. 3591). The Committee also noted the resolution (H. Res. 131) introduced by Representative Ottinger on January 19, 1977, requesting DHEW to regulate recombinant DNA research under Section 361 of the PHS Act.

Hearings held by State and local governments, including State legislatures, were among State and local activities reviewed. Recommendations for State regulation by the New York State Attorney General's Environmental Health Bureau, and for city regulation by the Cambridge (Massachusetts) City Council, were also considered.

Several committee representatives also reported on meetings with other interested parties whose views had been solicited on legislation to regulate recombinant DNA research. Those who were contacted included agricultural scientists, biomedical scientists, environmentalists, labor unions, and private industry. At the request of the Chairman of the Committee, the Industrial Research Institute and the Pharmaceutical Manufacturers Association are surveying their member firms to determine the scope of the research efforts in the private sector. The Pharmaceutical Manufacturers Association has adopted the NIH Guidelines for safe conduct of this research.

In light of this review, the full Committee recommends that the following elements should be included in proposed legislation for the regulation of recombinant DNA research:

(1) DEFINITIONS

"Recombinant DNA molecules" should be defined in a manner consistent with the NIH Guidelines.

Through an appropriate definition of the term "person," the legislation should cover any individual, corporation, association, Federal, State, or local institution or agency, or other legal entity.

"Secretary" should mean the Secretary of Health, Education, and Welfare.

(2) GENERAL REQUIREMENTS

The legislation should bar any person from engaging in the production or use of recombinant DNA molecules in a State of the United States, in the District of Columbia, the Commonwealth of Puerto Rico, the Virgin Islands, American Samoa, Guam, the Trust Territory of the Pacific Islands, Wake Island, Outer Continental Shelf Lands as defined in the Outer Continental Shelf Lands Act, Johnston Island, or the Canal Zone, unless (a) such production or use is permissible under standards promulgated by the Secretary, (b) such production or use is in compliance with any such standards, (c) the licensing requirements prescribed in the legislation have been satisfied, and (d) the registration requirements prescribed in the legislation have been satisfied.

The legislation should permit the Secretary to exempt activities from these requirements (a) where the activity is for specific commercial purposes found by the Secretary, after consultation with the regulating agency, to be regulated under other Federal law, or (b) where the Secretary determines that the activity poses no unreasonable risk to health or the environment.

(3) STANDARDS

The Secretary should be directed, as soon as practicable after passage of the legislation, to promulgate the NIH Guidelines for Research Involving Recombinant DNA Molecules as initial standards, with such clarifications and modifications as the Secretary determines to be necessary. Standards should assure, on the basis of the best currently available evidence, that no employee will suffer material impairment of health or functional capacity even if such employee engages in the production or use of recombinant DNA molecules for an entire working lifetime.

The legislation should authorize the Secretary to modify and revoke any of these initial standards and to promulgate new standards.

The legislation should include an appropriate provision for judicial review.

(4) LICENSURE OF LABORATORIES

The legislation should bar any person from engaging in the production or use of recombinant DNA molecules except at a facility licensed by the Secretary. A license should not be issued unless the Secretary determines that the facility will be operated in accordance with standards promulgated under the legislation and such other conditions as the Secretary deems appropriate.

The Secretary should have authority to exempt from the licensure requirement categories of activity which he determines pose no unreasonable risk to health or the environment. He should also, at his discretion, be able to utilize qualified accreditation or licensing bodies to assist him in carrying out this licensing function.

The legislation should have appropriate provisions for revocation, suspension, and limitation of licenses and for judicial review.

(5) REGISTRATION

The legislation should bar any person from engaging in the production or use of recombinant DNA molecules unless the activity has been registered with the Secretary, provided that the Secretary should be able to exempt from the provisions of this section categories of production or use which he determines pose no unreasonable risk to health or the environment.

(6) IMMINENT HAZARDS

The Secretary should have authority to sue to enjoin the production or use of recombinant DNA molecules where he believes the activity would constitute an imminent hazard to health or the environment.

(7) INSPECTIONS, SUBPOENAS, RECORD-KEEPING, AND REPORTS

The Secretary, in carrying out the legislation, should have authority to inspect facilities, make environmental measurements, conduct medical investigations, inspect medical records, issue subpoenas and citations, and require record-keeping and reports.

(8) DISCLOSURE OF INFORMATION

The legislation should provide that all records submitted to, or otherwise obtained by, the Secretary or his representatives under the legislation shall be available to the public upon request, except (a) information now exempt from disclosure under the Freedom of Information Act, and (b) other information the disclosure of which would cause the loss of proprietary rights.

At the time of request, persons who have submitted records should be given an opportunity to identify those portions which they believe to be excepted from disclosure under the preceding paragraph. The Secretary should not release such portions unless (a) he has found the portions so identified not to be excepted and has given the submitter advance notice of this finding and an opportunity to rebut it, or (b) the public need to know so outweighs the interest of the submitter as to require release. Where the Secretary releases records or portions thereof because of the public need to know, he should notify the submitter, setting forth the urgent health or environmental needs which serve as the basis for his action.

(9) COORDINATION

The legislation should provide specifically for interagency coordination in setting standards and avoiding duplicative requirements.

(10) PREEMPTION

The legislation should specifically preempt all State and local laws regulating the production or use of recombinant DNA molecules; except that where a State passes a law imposing requirements identical to those contained in the Federal legislation, the Secretary should have discretion to enter into an agreement with the State to carry out the Secretary's responsibilities under the legislation.

(11) ENFORCEMENT

The legislation should contain provisions for enforcement and sanctions.

(12) EMPLOYEE RIGHTS

The legislation should contain protections for employees who cooperate in the enforcement of these provisions.

(13) SUNSET

The legislation should remain in effect for a period of five years from the date of enactment, unless further action is taken by Congress.

VI. SUGGESTED ELEMENTS FOR LEGISLATION: COMMITTEE ANALYSIS

In considering these elements for proposed legislation, a number of issues were raised and discussed by the Committee. The issues that the Committee considered of importance are described below.

(1) DEFINITION OF THE TERM "SECRETARY"

The Committee considered the appropriate locus in the Government for the regulation of the use and production of recombinant DNA molecules. It determined that the Department of Health, Education, and Welfare is the appropriate locus in light of

(a) NIH's role as a lead agency in setting the standards,
(b) the petition by the Environmental Defense Fund to DHEW to issue regulations in this area,
(c) the congressional proposals that placed regulatory responsibility in DHEW, and
(d) the experiences of DHEW's Center for Disease Control in regulating infectious agents, and of its Bureau of Biologics (FDA) in licensing the production of biological products, in close cooperation with other Federal Departments and agencies.

This recommendation was formally approved by all members of the Committee. The Committee also urges close cooperation and coordination in DHEW between the NIH and regulatory agencies to ensure effective implementation of the standards set for this research.

(2) THE SCOPE OF REGULATION

The Committee reviewed at great length the nature and scope of regulation. Consideration was given to regulation of all laboratory research where hazardous or potentially hazardous substances were employed. Dr. Fredrickson reviewed the activities of committees at the NIH other than the Recombinant DNA Molecule Program Advisory Committee which have been created to study and recommend, if necessary, safety standards for other research involving actual or potential biohazards.

There was general Committee agreement that, for the present, legislation should be restricted to recombinant DNA techniques, allowing for sound administrative and scientific expertise in developing safety standards and regulation in other areas. The Committee considered whether, in the proposed legislation, the regulations should be limited to research. As noted above in the analysis of existing legislation, no current single, legal authority reaches all research under requirements set for regulation by the Committee. However, the Occupational Safety and Health Administration and the Environmental Protection Agency do have authority for regulation of commercial applications of recombinant DNA molecules.

Regulation of research alone presents a problem because of the difficulty in determining the border between research and pilot production. Therefore, the Committee recommends that regulation cover the production or use of recombinant DNA molecules. Such language would include research activity, and makes immaterial any consideration of whether a given activity constitutes research, pilot production, or manufacture. The Committee recommends that the Secretary, in consultation with appropriate regulatory agencies, be allowed to determine the nature of the activity and should defer to a regulatory body he determines is better empowered and equipped to deal with it.

The Committee also recommends as a suggested element for legislation a "sunset provision" for the regulatory authority. This provision is intended to mandate a review of regulation in light of accumulated scientific and safety information. This provision, the Committee wishes to emphasize, does not refer to records and other data relevant, for example, to medical, occupational, or environmental surveillance.

(3) REGISTRATION

There was general agreement by the Committee that registration of projects and other activities involving the use or production of recombinant DNA molecules was an important element of regulation. It was the consensus of the Committee that registration should occur prior to the initiation of the project, but that approval before commencing the project should not be required. Further, the Committee recommends that the Secretary have the authority to exempt certain classes of projects from this requirement.

(4) LICENSURE OF FACILITIES

It was the consensus of the Committee that the licensure provision should apply only to facilities, and that the facility would, under the terms of its license, accept responsibility for the particular activities and individuals at the facility. The Committee concluded that licensure of the facility and registration of projects would meet the needs for safety monitoring without extension of licensure to the projects themselves. The Committee discussed the possibility of revoking a license for serious and willful violations of the regulations. There was concern expressed that revocation was a very punitive measure, but it was agreed that the Secretary may wish to consider it for serious violations of the standards.

(5) DISCLOSURE OF INFORMATION

It was the scientific community that brought to public attention potential hazards of recombinant DNA research, and the NIH Guidelines, in that spirit, promote disclosure and dissemination of scientific and safety information. The Committee urges full disclosure to the appropriate regulatory body of all relevant safety and scientific information on the use or production of recombinant DNA molecules. However, the Committee recognizes the important world-wide commercial potential of recombinant DNA molecules in medicine, agriculture, and other areas of science and technology. It believes that the potential commercial uses of recombinant DNA techniques require that information of a proprietary nature and patent rights be given appropriate protection from disclosure by the regulatory agency receiving such information. Some Committee members concern that universities and inventors with limited resources may be unable to adequately protect data of a proprietary nature if the regulatory agency acts to disclose such information. The regulatory agency should consider the burden of its action on these inventors.

(6) PREEMPTION OF STATE/LOCAL LAWS

The potential hazards posed by the use of recombinant DNA techniques extend beyond the local to the national and international levels. Therefore, the Committee recommends that a single set of national standards must govern and that, accordingly, local law should be preempted to ensure national standards and regulations. The Committee, however, took into account the activities at the State and local levels on regulation of recombinant DNA research. It was agreed that if a State passes a law imposing requirements identical to those contained in the Federal legislation, then the Secretary may enter into an agreement with the State to utilize its resources to assist the Secretary in carrying out his duties.

(7) INSPECTION AND ENFORCEMENT

The Committee proposes that there be inspection and enforcement requirements to ensure that standards are being met. In order to protect the public health from an imminent potential hazard, the Committee also recommends that the Secretary have authority to enjoin the use or production of recombinant DNA molecules when he deems it necessary.

The Committee also reviewed the question of civil liability in the event of injury to humans or the environment. It believes that actions for damages should be left to State and local law. It is concerned that the inclusion of standards for strict liability as proposed in S. 621 could place a severe constraint on the ability of an institution to obtain liability insurance. It was predicted that, without insurance, institutions might have to terminate their research efforts unless national legislation were passed to indemnify them against adverse judgments.

(8) INTERAGENCY COOPERATION

Because of the wide potential use and production of recombinant DNA molecules and the need for uniform development and implementation of standards, the Committee recommends that mechanisms be established by the Secretary to ensure cooperation and coordination among appropriate Federal Departments and agencies. The National Institutes of Health is developing appropriate liaison between its Recombinant DNA Molecule Program Advisory Committee and relevant Federal research agencies, such as the Department of Agriculture, the National Science Foundation, and the Energy Research and Development Administration.

VII. FUTURE AGENDA

Pending action on possible legislation, the Committee stands ready to assist DHEW or whatever agency is made responsible for regulation of activities involving the use or production of recombinant DNA molecules. For example, research agencies on the Committee are working in coordination with the National Institutes of Health and its Recombinant DNA Molecule Program Advisory Committee on setting standards and certifying new host-vector systems. The research agencies have also been developing a registry of projects supported by Federal funds. The survey being taken in the private sector by the Pharmaceutical Manufacturers Association and the Industrial Research Institute will provide data on the industry, in anticipation of registration under a new law.

The Committee will consider suggestions by the representatives from the State Department concerning further means to ensure international control in the use and production of recombinant DNA molecules. At present, there is voluntary coordination and cooperation among national scientific bodies. The Biological Weapons Convention is considered by the State Department to prohibit development, production, or stockpiling of recombinant DNA molecules for purposes of biological warfare. The Committee will review whether other measures need to be considered for international control.

The Committee will also be reviewing current Federal policies on the matter of patenting recombinant DNA inventions and other matters of concern that may need to be addressed before the Committee concludes its business and files a final report.

Report of the Interagency Committee on Recombinant DNA Research: International Activities, November 1977

An introductory summary and Part A, "Committee Analysis and Recommendations," have been reprinted here. Part B contains detailed information on the activities of scientific organizations in individual countries. For information on Part B and the appendices, contact the Office of the Associate Director for Program Planning and Evaluation, National Institutes of Health, Bethesda, Md. 20014.

SUMMARY

On May 6, 1977, the Federal Interagency Committee on Recombinant DNA Research met to consider possible means for fostering international control of the use and production of recombinant DNA molecules. A Subcommittee on International Issues was created. It met in June and July 1977 to review international activities in this field with a view to making recommendations on means for achieving common safety standards. By way of assistance, NIH staff surveyed all recombinant DNA activities occurring internationally, and the State Department had the survey reviewed by U.S. science attachés abroad.

The Subcommittee's analysis is presented. It reveals that scientists throughout the world have played a leading role in bringing the potential hazards of recombinant DNA research to the attention of scientists, governments, and international organizations. As a result, efforts have been made to adopt safety procedures for the conduct of the research in many countries. The NIH and U.K. guidelines, imposing similar safeguards, are being

used as important models. The United States has about 300 active projects involving the new technique, and about half that number are under way in Europe. All are being done under some form of safety standards.

National and international safety organizations in many countries have studied the recombinant DNA issue, usually recommending some form of control. The European Molecular Biology Organization, the International Council of Scientific Unions, and the World Health Organization endorse both the NIH and U.K. Guidelines, and the European Science Foundation endorses the latter. The activities of these and other organizations are described.

International organizations are also examining the ethical implications of biological research generally and genetic research particularly. UNESCO held meetings to this end in 1975 and October 1977, and WHO and the Nobel Foundation will sponsor conferences in 1978 to examine ethical, social, and legal questions. In the United States, activities have been undertaken to address similar ethical concerns. Appropriate government agencies are expected to study these important issues.

The Federal Interagency Committee recommends that U.S. Government agencies continue to work closely with national, international, and regional organizations to promote safeguards and disseminate relevant information. In the Committee's view, no formal governmental action is necessary at present to produce international control by means of a treaty or convention. The Committee emphasizes that the Biological Weapons Convention prohibits the use of recombinant DNA for biological warfare.

Results of the survey are presented alphabetically by country. The nations comprising international organizations are listed. Nine appendices contain key documents, including the U.K. Guidelines.

I. INTRODUCTION

In September 1976 a Federal Interagency Committee on Recombinant DNA Research was created to consider an extension of the "NIH Guidelines for Research Involving Recombinant DNA Molecules" beyond the NIH to the public and private sectors. The Committee was convened by the Secretary of Health, Education, and Welfare (HEW) with the approval of the President. Dr. Donald S. Fredrickson, Director of the National Institutes of Health (NIH), serves as chairman at the Secretary's request. (See Appendix I for Committee membership.)

The Committee met eight times from November 1976 to March 1977. Details of these meetings, and the recommendations of the Committee for legislation to govern the conduct of recombinant DNA activities nationally, are contained in the Committee's Interim Report, released by the Secretary of HEW on March 16.*

After making its recommendations for legislation, the Committee met, at the request of representatives from the State Department, to consider possible means for fostering international control of the use and production of recombinant DNA molecules. Dr. Fredrickson reported on the voluntary coordination and cooperation that exist among national and international scientific bodies to promote uniform safety practices and procedures. Further, the representative from the U.S. Arms Control and Disarmament Agency reported that the State Department considers the Biological Weapons Convention as prohibiting development, production, or stockpiling of recombinant DNA

---

*Copies of the Interim Report may be obtained from the Office of the Associate Director for Program Planning and Evaluation, National Institutes of Health, Bethesda, Maryland 20014. (301) 496-3152.

molecules for purposes of biological warfare. It was agreed, however, that the Committee, prior to concluding its business, should inquire into the activities in other countries relative to the use of recombinant DNA techniques and should make recommendations to the Secretary of HEW regarding Federal measures to promote international control. Recombinant DNA research has aroused widespread international debate, and the actions in other countries deserve careful attention when U.S. policy is being developed for legislation and regulation. (See Appendix III.)

Accordingly, a Subcommittee on International Issues of the Interagency Committee on Recombinant DNA Research was created for the purpose of preparing an analysis of international activities to date and recommending means for achieving common safety standards wherever the use of recombinant DNA techniques takes place. The work of the Subcommittee and its recommendations are the basis for this Committee report. (See Appendix II for membership of the Subcommittee.)

II. SUBCOMMITTEE REVIEW OF INTERNATIONAL RECOMBINANT DNA ACTIVITIES

Two meetings of the Subcommittee on International Issues were held to review recombinant DNA research activities outside the United States. To assist in this review, NIH staff prepared a survey of all recombinant DNA activities occurring internationally. The State Department subsequently sent the survey to U.S. science attachés abroad, and it was revised in response to comments received. The survey appears as Part B of this report.

The Subcommittee also received reports from a number of persons who serve in or work with various international organizations involved in recombinant DNA research activities. On the basis of an analysis of this information, the Subcommittee developed several recommendations that were transmitted to the full Committee for consideration. The analysis and recommendations follow.

III. INTERNATIONAL SCIENTIFIC ACTIVITIES: AN ANALYSIS

Scientists in the United States and abroad have played a leading role in bringing the potential hazards of recombinant DNA research to the attention of scientists, governments, and international organizations. As a result, safety procedures for the conduct of this research are being adopted in many countries. Although nations differ in their perceptions of the need to adopt safety measures, and of the nature of these measures, the NIH and

U.K. Guidelines are being used as important models. (See Appendix IV for the "Williams Report," also known as the "U.K. Guidelines.") As of the summer of 1977, there were an estimated 150 recombinant DNA projects under way in Europe, 300 in the United States, and perhaps 20–25 altogether in Canada, Australia, Japan, and the Soviet Union. All are being conducted under some form of safety practices and procedures.

Details concerning the activities of individual scientific organizations are given in Part B.

A. NATIONAL SCIENTIFIC ORGANIZATIONS

The issue of recombinant DNA research has been studied by national and international bodies in countries throughout the world. In many cases some form of control has been recommended, but in no case has a total ban on the research been advocated. The United Kingdom and Canada have issued guidelines that differ in detail but are similar conceptually to the NIH Guidelines.* Other countries are generally following the NIH or U.K. Guidelines, including Denmark, the Netherlands, the German Federal Republic, Israel, Sweden, and Switzerland. Endorsement of the U.K. Guidelines has been given by the European Science Foundation (ESF). The European Molecular Biology Organization (EMBO) endorsed the use of either the U.K. or NIH Guidelines. The International Council of Scientific Unions (ICSU) and the World Health Organizations (WHO) have urged nations to adopt the principles embodied in these two sets of guidelines.

As detailed in the international survey, scientific and governmental activities comparable to those in the United States have been under way in the United Kingdom since July 1974. A working party established at that time recommended that recombinant DNA research in the United Kingdom be permitted to continue under appropriate controls. A followup working group chaired by Sir Robert Williams issued a report in August 1976 that established guidelines for the conduct of recombinant DNA research within the United Kingdom.

In Canada, in March 1976, a special committee of the Canadian Medical Research Council recommended guidelines to govern the handling of recombinant DNA molecules in Council-supported research. The Council adopted these guidelines in February 1977. Many other nations have also reviewed recombinant DNA activities to determine the measures necessary for safety. With

the urging of regional and international bodies, however, most have adopted the NIH or U.K. Guidelines as a basic framework for safety practices and procedures.

B. REGIONAL SCIENTIFIC ORGANIZATIONS

The European Molecular Biology Organization, the European Science Foundation, and the European Medical Research Councils (EMRC) have been instrumental in closely coordinating recombinant DNA research activities in western Europe. All three have worked closely to promote a commonality of safety practices and procedures to govern recombinant DNA research activities. With the support of these organizations, scientific technical committees have been created in the western European countries to serve as a focus for coordinating and monitoring all recombinant DNA activities within each country. Many of these bodies function in the manner of the United Kingdom's Genetic Manipulation Advisory Group, which is responsible for reviewing all recombinant DNA research to ensure that the projects conform to appropriate safety standards and practices.

Research is also being conducted in the eastern European countries and the Soviet Union. They are considering the adoption of safety practices comparable to those specified in the U.K. or NIH Guidelines. The U.K. or NIH standards serve as a model to govern recombinant DNA research under way in Japan and Australia. Other than in Japan, there is little known activity in Asian, African, or Latin American countries.

C. INTERNATIONAL SCIENTIFIC ORGANIZATIONS

International scientific organizations involved in recombinant DNA activities include ICSU and WHO. The activities of WHO are described under health and safety efforts (section IV). A description of the activities of ICSU and its Committee on Genetic Experimentation (COGENE) was presented to the Interagency Subcommittee by Dr. William J. Whelan, Chairman of COGENE and Professor and Chairman of Biochemistry at the University of Miami School of Medicine. (Additional information about COGENE is given in Appendix V.) COGENE was established in October 1976 and has created three task forces to study and analyze various aspects of recombinant DNA research. These groups and their current tasks are as follows:

*A Working Group on Recombinant DNA Guidelines, chaired by Dr. S. Cohen of the United

---

*Copies of the Canadian guidelines may be obtained from the Director, Special Program, Medical Research Council, Ottawa, Canada K1A OW9.

States, is reviewing extant guidelines to determine important similarities and differences.

*A Working Group on Training and Education, chaired by Professor K. Murray of the United Kingdom, is seeking in its efforts the assistance of the United Nations Environmental Program (UNEP) and the United Nations Educational, Scientific, and Cultural Organization (UNESCO).

*A Working Group on Risk Assessment, chaired by Dr. A. Skalka of the United States, is drafting a questionnaire surveying all relevant activities internationally.

COGENE also plans to sponsor, if support is available, an international conference for industrial representatives to discuss recombinant DNA activities in the private sector throughout the world.

COGENE has liaison representatives from the major European science organizations and relevant U.N. organizations, including UNEP, Food and Agriculture Organizations (FAO), UNESCO, and WHO.

IV. INTERNATIONAL HEALTH AND SAFETY ACTIVITIES: AN ANALYSIS

A. NATIONAL

A number of nations have initiated activities to foster government monitoring of recombinant DNA research for purposes of safety and health. In the United States the recommendations for legislation by the Interagency Committee were designed to allow this significant research to continue under uniform national safety standards that would assure the best possible protection for man and the environment from the effects of potential hazards, the nature of which are as yet unknown.

In the United Kingdom the government's Health and Safety Executive (HSE) will have the responsibility after October 1978 of ensuring that the standards of the U.K. Genetic Manipulation Advisory Group (GMAG) are followed. HSE has authority under the Health and Safety at Work Act to monitor the workplaces of all facilities. It maintains a force of inspectors for this purpose who have the authority to investigate industrial, governmental, and university laboratories. Advisory Groups similar to the U.K. GMAG have also been established in other European countries, and efforts are under way to identify appropriate governmental bodies to ensure compliance with safety standards for this research.

B. REGIONAL

The European Medical Research Councils, the European Molecular Biology Organization, and the European Science Foundation strive to ensure consistency in safety standards to govern this research throughout western Europe. All have varying degrees of contact with government organizations to enlist their assistance and to coordinate activities. Under the sponsorship of NIH and EMBO, a workshop was convened in London in March 1977 to discuss parameters of physical containment.* The technical aspects of primary and secondary physical containment were discussed, and agreement was reached on the appropriate range in which physical containment provisions should fall. The report of this meeting indicates that, while adjustments are still required if these guidelines are to offer uniform physical containment standards, all three sets are comparable. This effort complements the work of WHO as described below (section IV-C).

The European Economic Community (EEC), as discussed in the survey, has legal authority under certain circumstances to enact policy decisions binding on its member nations. In this context, EEC has begun to examine scientific activities of member states to ensure that safety measures they have adopted are uniform and that private industry adheres to the same standards as the public sector. Meetings have been held, and a directive is currently under consideration which would require each member state to establish its own administrative mechanism to ensure that all recombinant DNA research is subject to national guidelines.

The application of safety standards to the private sector raises policy questions concerning the international protection of proprietary information and patent rights. At a meeting convened in Strasbourg in March 1977 under the auspices of ESF, the problem of disclosure of information was reviewed as it relates to the operations of the various national committees of the member countries. The discussion centered on balancing the need for disclosure of information in research protocols among nations (in order to implement uniform safety practices) with the need to protect proprietary information and patent rights. As a result, a meeting of European patent experts was convened in June for the purpose of studying the complex issues of proprietary rights and patent laws.

---

*The "Report of the NIH/EMBO Workshop (Parameters of Physical Containment)" may be obtained from the Office of Research Safety, National Cancer Institute, Room 2E47, Building 13, National Institutes of Health, 9000 Rockville Pike, Bethesda, Maryland 20014.

Another organization, the World Intellectual Property Organization (WIPO), has drafted a treaty and regulations pertaining to the international recognition of the deposit of microorganisms for purposes of patent procedures. (See summary of the treaty, Appendix IX.) WIPO is an independent organization, funded through grants and contracts from a number of private and public entities, which has members of various scientific societies on its board of directors. A final treaty is under negotiation and would presumably speak to some of the needs that have arisen in the context of recombinant DNA inventions.

As previously mentioned, COGENE (of the International Council of Scientific Unions) is seeking financial support to convene a meeting of international representatives from the private sector to consider the questions of patents and the commercial use of this technology.

C. INTERNATIONAL

The World Health Organization has initiated a number of activities to promote safety practices in recombinant DNA research. In June 1975 the Advisory Committee on Medical Research (ACMR) of WHO issued a report concluding that recombinant DNA research should continue under appropriate safeguards. In September 1976, in Geneva, NIH and WHO jointly organized a conference on the subject of safety practices for the international transfer of research materials, especially in respect to infectious substances. As a result of that conference, it was recommended that WHO establish an Advisory Group for Safety Measures in Microbiology and that four international working groups be created to initiate and monitor activities in the following areas: safe transfer of infectious materials, laboratory safety clements, maximum containment laboratories, and development of emergency services.

ACMR accepted these recommendations in June 1977. Dr. K. Bögel currently heads the Special Programme on Safety Measures in Microbiology; the four working groups have been created; and NIH is considering support for this program to develop international biological safety standards. (See Appendix VI.)

At a meeting in June 1977, a subcommittee of ACMR recommended that WHO have a role in disseminating information and assisting other scientific organizations in performing risk-assessment studies relating to recombinant DNA research. It was further recommended, however, that the technical analysis of guidelines, the performance of risk-assessment studies, and other technical activities be left to organizations such as ICSU. These recommendations were made in light of the close

cooperation and coordination among international scientific, health, and safety organizations.

For March 1978 WHO has also scheduled a symposium entitled "Practical Applications of Genetic Engineering." This will provide a forum for discussion of the potential application of benefits from this research, especially with regard to the lesser-developed countries.

V. ETHICAL IMPLICATIONS OF GENETIC RESEARCH

In light of the international interests in recombinant DNA research, international organizations have sponsored a number of conferences to examine the ethical implications of biological research generally and genetic research particularly. These include the following:

* UNESCO sponsored a meeting in Bulgaria, in June 1975, to examine the ethical, legal, and social implications of research advances in biology, including advances in genetic research;
* UNESCO sponsored a meeting in Madrid, Spain, in October 1977, to examine the ethical implications of genetic research, including recombinant DNA;
* WHO will sponsor a meeting in Milan, Italy, in March 1978, to examine the implications of genetic research for "genetic engineering"; and
* the Nobel Foundation will sponsor a meeting in Stockholm, Sweden, in August 1978 to examine ethical questions in key science policy areas.

In the United States, activities have been undertaken to address similar ethical concerns. In both House and Senate hearings on recombinant DNA research legislation, questions have been raised by Committee members regarding the long-range ethical implications of this research. Indeed, in both House and Senate bills, there are provisions calling for studies concerning the broad social, legal, and ethical aspects of this research. Further study of these aspects will undoubtedly be initiated by appropriate government agencies.

Still other questions have been raised concerning the possible use of this research for biological warfare. With regard to this question, the use of recombinant DNA for such purposes is prohibited by the Biological Weapons Convention. In a statement to the conference of the Committee of Disarmament on August 17, 1976, Ambassador Joseph Martin, Jr., confirmed the U.S. interpretation of the convention to prohibit this research for purposes of biological warfare. (See Appendix VII for Ambassador Martin's discussion.)

It is noteworthy that prior to this statement, Dr. David Baltimore had requested an opinion from

James L. Malone, General Counsel of the U.S. Arms Control and Disarmament Agency, on whether the Biological Weapons Convention prohibits production of recombinant DNA molecules for purposes of constructing biological weapons. On July 3, 1975, Mr. Malone replied: "In our opinion the answer is in the affirmative. The use of recombinant DNA molecules for such purposes clearly falls within the scope of the convention's provisions."

VI. COMMITTEE RECOMMENDATIONS

In developing the following recommendations, the Interagency Committee on Recombinant DNA Research considered the activities to date of national, regional, and international scientific organizations and governmental health and safety organizations. Also, a concerted effort was made to obtain the most comprehensive possible survey of international activities related to recombinant DNA research. The survey of activities, prepared by NIH staff, was circulated by the State Department to all U.S. science attachés abroad, with a request to update the report and to supply any information available on governmental actions taken to legislate or regulate recombinant DNA activities. The survey, which appears as Part B, includes all relevant information received by the State Department.

Professor William J. Whelan, Chairman of the Committee on Genetic Experimentation of the International Council of Scientific Unions, attended both meetings of the Subcommittee and briefed it on the actions of his Committee and of other international scientific organizations. Further, a number of NIH officials reported on extensive NIH activities of interrelating U.S. efforts with those of WHO and European organizations such as the European Molecular Biology Organization, the European Science Foundation, and the European Medical Research Councils. After review and discussion of all the information obtained, the following recommendations are made:

* In light of the activities of national, regional, and international organizations, the Committee recommends that cognizant Federal agencies continue to work in close cooperation with such organizations as ESF, EMBO, EMRC, WHO, and ICSU. The Committee takes special note of the activities of ICSU and recommends that Federal agencies work with ICSU and support its working groups on recombinant DNA guidelines, training/education, and risk-assessment.

* The Committee recognizes the special need to ensure wide dissemination of information on all international recombinant DNA activities. Hence, the Committee recommends that relevant Federal agencies, such as NIH, the Department of Agriculture, and the National Science Foundation, conduct or support information programs to ensure that international developments in recombinant DNA activities are periodically and widely reported. It was noted that NIH's *Recombinant DNA Technical Bulletin* (formerly *Nucleic Acid Recombinant Scientific Memoranda*) is designed to serve as a useful vehicle in this regard.

* The Committee strongly endorses WHO efforts to achieve uniformity in safety practices in biological research. Therefore, it recommends that relevant Federal agencies provide support to WHO and other international organizations with the capability to develop international biological safety codes of practice.

* The Committee recognizes the importance of private-sector research activities and the problems presented in the protection of proprietary information and patent rights. The Committee, therefore, urges relevant Federal agencies, such as the Department of Commerce, to assist international efforts in seeking a resolution of these issues. The Committee further believes that the conference proposed by COGENE for private industry is of key importance and that support for the conference should be considered by relevant industries within the United States as well as by companies abroad.

* The Committee notes that a number of international organizations have held or are planning meetings to examine the social, ethical, and legal implications of biological research. The Committee urges that relevant Federal agencies closely monitor international activities addressing the ethical implications of genetic research (including techniques employing recombinant DNA) in relation to the broad subject of "genetic engineering."

* In noting the extensive international cooperation among scientific, environmental, health, and safety organizations, it is the Committee's view that no formal governmental action is necessary at present to produce international control by means of a treaty or convention. Such activities, in this Committee's view, would be premature. More international collaboration, especially among national health and safety organizations, is necessary to determine whether formal control at the international level is warranted and feasible. Finally, the Committee emphasizes that use of recombinant DNA for purposes of biological warfare is prohibited by the Biological Weapons Convention.

Additional Resources

(1) "Ethical and Scientific Issues Posed by Human Uses of Molecular Genetics." *Ann. N.Y. Acad. Sciences,* Vol. 265, 1976 (entire issue).

(2) *Final Environmental Impact Statement on NIH Guidelines for Research Involving Recombinant DNA Molecules of June 23, 1976.* Parts One and Two. DHEW Public No. (NIH) 1489 and 1490. For information contact the Director, National Institutes of Health, 9000 Rockville Pike, Bethesda, Md. 20014.

(3) *Genetic Engineering, Human Genetics and Cell Biology: Evolution of Technological Issues.* Report prepared for the Subcommittee on Science, Research and Technology of the Committee on Science and Technology, U.S. House of Representatives, Ninety-fourth Congress, Washington, D.C.: U.S. Government Printing Office, 1976.

(4) *Institute Archives of the Massachusetts Institute of Technology.* Oral History Collection, MIT 20B-231, Cambridge, Mass. 02139.

(5) *Recombinant DNA: Readings from Scientific American.* With introductions by David Freifelder. San Francisco: W. H. Freeman and Company, 1978.

(6) *Recombinant DNA Research.* Documents relating to "NIH Guidelines for Research Involving Recombinant DNA Molecules." Vol. 1, February 1975–June 1976; Vol. 2, June 1976–November 1977. For information contact the Director, National Institutes of Health, 9000 Rockville Pike, Bethesda, Md. 20014.

(7) *Recombinant DNA Technical Bulletin.* Published quarterly by the Office of Recombinant DNA Activities, National Institute of General Medical Sciences, Room 4A52, National Institutes of Health, Bethesda, Md. 20014.

(8) *Regulation of Recombinant DNA Research.* Hearings before the Subcommittee on Science, Technology and Space of the Committee on Commerce, Science and Transportation, U.S. Senate, Ninety-fifth Congress. Washington, D.C.: U.S. Government Printing Office, 1978.

(9) *Research with Recombinant DNA.* An Academy Forum, March 7–9, 1977. Washington, D.C.: Nat. Acad. Sciences, 1977.

(10) *Science Policy Implications of DNA Recombinant Molecule Research.* Hearings before the Subcommittee on Science, Research and Technology of the Committee on Science and Technology, U.S. House of Representatives, Ninety-fifth Congress. Washington, D.C.: U.S. Government Printing Office, 1977.

(11) *Science Policy Implications of DNA Recombinant Molecule Research.* Report prepared by the Subcommittee on Science, Research and Technology of the Committee on Science and Technology, U.S. House of Representatives, Ninety-fifth Congress. Washington, D.C.: U.S. Government Printing Office, 1978.

Glossary

Bacterifacture A coined term designating the production of desired materials by bacteria, as by instruction through inserted recombinant DNA.

Bacteriophages Viruses that reproduce within bacterial cells and infect new cells following the breaking open of the old ones.

Conjugation Exchange of DNA between two cells that are temporarily joined by a cytoplasmic bridge.

DNA replication The process by which a DNA molecule gives rise to two identical copies of itself.

Genome The sum total of hereditary information in a given cell or organism. Similar to genotype.

Genotype The herditary content that determines the genetic behavior of a given cell or organism. Now known to consist of the nucleotide sequence of the total replicated DNA.

Germ cells Cells produced and set aside early in development to produce sex cells or gametes. In higher animals only changes produced in germ cells or their descendants can be transmitted to the next generation.

Hereditary continuity The transmission, between generations, of replicative genetic determinants that make for likeness.

Hereditary variation Changes in replicative genetic developments that introduce new characteristics subsequently transmitted to descendant generations.

Host A cell or organism within which an introduced replicative agent can propagate itself.

Host-vector system A combination of a host and a vector that allows a replicative agent to be propagated, for example, a segment of foreign DNA.

Messenger RNA An RNA produced from DNA by transcription and able to be translated into the complementary amino acid sequence of a protein.

Nucleotides The molecular subunits that are linked in a chain to form nucleic acids, either DNA or RNA.

Nucleases Enzymes (proteins) that act to break the links between nucleotides in a nucleic acid chain.

Operon A genetic component including several genes and the control elements that determine when they will be expressed.

Phenotype The characteristics of a given organism resulting from the expression of its genotype in development under particular environmental conditions.

Plasmid A particle consisting of a relatively small, circular DNA molecule that replicates in bacteria independently of the larger chromosome.

Procaryote cell The less complex cell type found in bacteria and blue-green algae. To be distinguished from eucaryote cells, containing a membrane-bounded nucleus and found in all other organisms.

Restriction enzymes Nucleases that break the links between nucleotides at a particular point determined by a particular nucleotide sequence.

Reverse transcriptase An enzyme (protein) that promotes production of complementary RNA from DNA.

Transcription The production from DNA of an RNA with a complementary nucleotide sequence.

Vector A vehicle for transmission of a replicative agent from an infected to a noninfected host. In molecular genetics, specifically a plasmid or virus that can carry a foreign DNA into a host cell.

173

Index